The Power of Time

: From Einstein with Time

Dr. Poramest Boonsri

SEVEN Press
645/82 Soi Issarapub 9, Issarapub Road,
Hirunrugee, Thonburi, Bangkok, Thailand 10600
Tel. 668-5071-8587 or 662-466-8128

(untolddiscovery@gmail.com or marichatporamest@gmail.com)

Printed in the United States of America

Publisher's Cataloging-in-Publication data
Boonsri, Poramest.
The Power of Time: From Einstein with Time/ Poramest Boonsri.
p. cm.
Includes bibliographical references and index

ISBN :0998838608
1. The physics of time—Popular science —Non Fiction. 2. The equation of all equations, the time vectors and the time pyramid—New findings — Popular Science. 3. Discovery in physics in 1987: the formation of magnetic poles—Non Fiction—Thailand. 4. New characteristics of Time—New finding —Popular Science. 6. Controlling time—Experiments—Non Fiction. 7. New Economics—linkage between micro-and macroeconomics—Non Fiction.

I.   II. Title.

First Edition

0 9 8 7 6 5 4 3 2 1

Preface

This book is about the next characteristics of time after Einstein found time to be a relative value in 1905. Albert Einstein was the first scientist who found the first characteristic about time; thereby, developed his well-known equation, $E = mc^2$, after he had found that time has special characteristic from the Maxwell's equation. Scientists have tested his part of the theory about time by using two atomic clocks put in the different places; they found the two clocks running with the different speeds and accepted the accuracy of Einstein's prediction. Time has unique characteristic just like some elements in the periodic table. After that, in 1988, Stephen Hawking wrote his book about time; where he referred to some aspect of time as the "arrow of time" but it was not new characteristic of time. In 2002, Paul Davis wrote his book "About time: Einstein's Unfinished Revolution" to present many views about time; however, the book did not present any new characteristic of time.

In 1987, I discovered something about magnetic field at Khon Kean University in Thailand. At that time, not even my scientists' friends and I could provide answers to what it was. Only two aspects that I noticed from the magnetic fields are that it was about time and a pyramidal shape. When I started to study about time, I really realized that time will be the dimension that we could not overlook anymore.

In this book, the next characteristics of time were found and presented by mathematically proving and explanation with popular science for readers who want to know what will happen in the future of science in words that we use every day. Values of time that match with the positions of numbers in number lines and which can be calculated, are second characteristic of time after Einstein's work. No one has ever described time in terms of numbers. In this book, time is described as everything because mathematicians use numbers representing everything. Further, time is presented as a vector, which implies that time has distinct directions. The direction of time is so important that we could not overlook it. The geometry of the time dimension that showed four main angles in three-dimensional body opened the secret of time in subatomic level. Scientists know the magic angle 54.7356° in subatomic level; they even know how get the angle through calculation. However, what they do not know is why the angle exists in subatomic level. Time in magic angle is scientific evidence supporting existence of time in subatomic level. It gives us new understanding of the pyramidal shape; that is, the shape of a time-field with monopole which links finally to a pole of a magnet.

The new characteristics of time are discussed in the book and the unseen results in science and social science presented as follow:

(1) in physics, **gravity force** has its poles. The polarity of gravity will solve the energy crisis in the middle of this century for people. It also solves the problem of transportation and will help physicists to solve their theoretical problem about the unity of four fundamental forces in nature into one coherent picture. **(2) in physics**, time opens one of the greatest unsolved mysteries in physics, the interaction of two 3-D bodies. Its application can help us in judging rays of light as a wave or a particle at the same times while two photons are spinning. It also indicated my discovery in 1987 as the formation of external magnetic poles that everybody has never seen before. We know magnet having its poles, but no one knows how the poles are formed. The linkage between quantum theory and Einstein's theory working well together may begin at this point, **(3) in chemistry**, time told us that scientists can control the growth of wanted food cells and unwanted cancer cells and can produce free energy for people to use in the future. The conclusion came from experiments of Thai scientists presented in an international journal in Thailand, **(4) in mathematics**, Michio Kaku, who mentions on his website about the one-inch equation that can explain all forces in physics but nobody had found until now. The time equation shows that the existence of an equation explaining many equations in physics is real and well-founded, **(5) in mathematics**, the time equation indicated that mathematicians can solve N equations with a value of time at one time. This finding will lead us to invent “air screen” of computers with very low prices. Advanced robots calculating with super speed will be developed directly for future of technology of robot, **(6) in economics**, time can link micro- and macroeconomics together. The two levels of economic study that have never worked together since Keynes created his macroeconomic theories eighty years ago were solved because of the new characteristic of the time. The new price model and Exchanging Line can describe how micro- and macroeconomics work well together. Time and Keynes’ theory can point out the new views of solving economy problems in the real world. The real cases of economic problems in Los Angeles were used to show how new price model solve problems. Time helps economic theories to be developed to the new branch of science. Moreover, time, the new production theory, and the wealth of nation were discussed. Readers who may want to know what will happen in their economy can see their future from this book. The projects of H.M the King Rama IX of Thai people who can develop the economy of Thailand from underdevelopment to developing level with high income were used to show how new price model works.

Something different from your expectation of future is waiting for your reading of this book. Another path of future science and social science is in your hand to choose now.

Acknowledgement

Thanks to Apichai Sivapraphagorn and his team, Aranya Pimmongkol, Warinee Parasarn, and Supon Samran, for his sincere support in various aspects, especially their experiments and many pictures drawn and presented this book. Special thanks to Andre′ Michaud, one of the talented physicists whom I have known for his guidance and correction of my English in some part of this book. Thanks to the mathematician, Associate Professor Kumjorn Muneekaew. Thanks to the love and the support of my king, idol and family who always give me big smiles. Special thanks to Schlastica, rinzmann and hmd_gfx at www.Fiverr.com who wrote my work in native English, did interior design and did the cover of my book respectively. Special thanks to Professor Dr. Apichai Puntasen who did my book review.

Table of Contents

Tables

Figures

Chapter 1
The Physics of Time

What we know is a drop, what we don't know is an ocean.

Isaac Newton

The position of time value is one unique concept that indirectly concerns the paradigm of quantum mechanics theory; however, it directly concerns the paradigm of Einstein's theory. Both theories are useful in explaining the two worlds of physics. Notably, whereas Quantum Theory is used to explain the tiny world in subatomic level, the Einstein's theory is used to explain the celestial world of the universe. Despite the successful application of the two theories in their unique contexts, they cannot be used to explain each other. As a result, there are various complexities in appreciating the nature of time when the two concepts are simultaneously used.

We are aware of the composition of small nature, which range from the tiny world in subatomic level to the big world like the solar system in a galaxy such as the universe. Consequently, there is uniqueness in the nature of both the small and big worlds; nevertheless, they are inseparable. Various physicists believe that in the future, there will be one distinct theory that can best explain the whole nature since the universe is not separated by the existing theories in physics. Thus, there is need to appreciate the two theories, especially in the context of time dimensions.

A brief overview of quantum mechanics theory

The difference between the quantum mechanics theory and the Einstein's theory is that the former successfully explains the tiny world in subatomic level while the latter explains the enormous world especially gravity and light. The concept of quantum mechanics theory was integrated in physics in 20th century, where different physicists tried to solve the question about black body. Questions concerning the black body revolved on issues such as "an iron bar used for making a sword can show light when it is very hot". The existing theories of physics at that time followed the Newton's mechanics theory to try to explain problems about the black body; however, not even one theory could successfully explain the whole idea.

Before the 20th Century, physicists tried investigating the structure of an atom. Consequently, they tried to preset the atomic models that could be used to explain the questions about the atom. The models varied over time due to the continued finding of new particles. For example, Sir Joseph John Thomson (1856-1940) explained the model of the atom that looked like the earth and the moon. The positive particles and many electrons were orbiting around the nucleus of the atom ("Sir Joseph John Thomson - Dictionary definition of Sir

Joseph John Thomson | Encyclopedia.com: FREE online dictionary", 2016). Having established new evidence from the experiments about neutron in the atom, Lord Ernest Rutherford (1871-1937) explained a new model of the atom that positive particles and neutral particles called neutrons in the nucleus of the atom and electrons are orbiting around the nucleus ("Rutherford, Ernest – Biography – Te Ara Encyclopedia of New Zealand", 2016).

There are other important equations that have been used to explain the concept of black body. Mark Plank discovered an equation known as the Plank's equation, which is used to explain the radiation of a black body. The equation was discovered after his guest named Heinrich Rubens had reported his last experimental result of the black body to Plank. The radiation was emitted discontinuously that was called "quanta." It depended on frequency of electromagnetic wave. Quantum mechanics theory was called by physicists at that time because of this reason ("Max Planck - Biographical", 2016; Parker, 1987).

Because of Plank's equation, models of the atoms were continuously developed. For example, Neils Bohr explained that the atom looked like two electrons orbiting around the nucleus with mathematics called Balmer Series. In addition, how quanta of energy were made were explained as well. The model could not explain the complicated atoms, which had three or more electrons ("Niels Bohr - New World Encyclopedia", 2016; Parker, 1987).

Louis Victor De Broglie explained that light or photon had two characteristics: mass and wave ("Louis de Broglie - Biography, Facts and Pictures", 2016; Parker, 1987). Schrodinger (1887-1961) discovered Schrodinger's equation for explaining of Broglie's model in different views; the equation explained the characteristic of particles as wave that was called Wave function Psi. This equation was one of principles of quantum theory ("Erwin Schrodinger facts, information, pictures | Encyclopedia.com articles about Erwin Schrodinger", 2016). The model with the equation explained electrons in the atom looked like cloud called "electron cloud" that could expand itself bigger and bigger.

Max Born (1882-1970) analyzed the Wave Function Psi, whereby, he described it as a wave of probability where electrons are found within the atom at a particular area. The views of the atom changed from the exact positions of electrons in the atom, with various speculations on the possible existence of electron positions in the atom. This analysis supported the work of Heisenberg (1901-1976) ("Max Born - Biographical", 2016; "Werner Heisenberg - Biographical", 2016; Parker, 1987).

Einstein (1879–1955) disagreed with the concept of possibility of finding electrons in the atom; he articulated that the quantum mechanics theory brought about unpredictability to the core of the theory. Additionally, Einstein indicated the subatomic world and the gigantic world as the universe do not occur based on the probability or coincident. Einstein said that God did not play disc; however, a little physicists heard his commend because of the success of quantum theory ("Albert Einstein | German-American physicist", 2016; Kaku and Thompson, 1987).

Paul Adrien Maurice Dirac (1902-1984) found that Schrodinger's equation and Einstein Relativity Theory could not run smoothly together; he created a new equation called "Dirac Relativistic Wave Equation" to study very high speed particles. The model of the atom was more when the details of characteristic of spin of particles were added. With the equation, Dirac concluded that there were twin of electrons with positive charge called "positron" (Parker, 1987; "Paul Adrien Maurice Dirac facts, information, pictures | Encyclopedia.com articles about Paul Adrien Maurice Dirac", 2016).

Quantum theory can be used to explain many phenomena in the atom; such as, why metals in general having high electrical conductivity, high thermal conductivity, and high density? With quantum theory, a semiconductor material that has an electrical conductivity value between a conductor like copper and an insulator such as glass, can be explained by the movement of electrons and holes in a crystal lattice

("Semiconductor - New World Encyclopedia", 2016). The knowledge of semiconductor materials and fabrication processes help scientists in making speed of microprocessors and memory devices with higher complexity. Besides, they are used on daily basis in the computers (Kaku and Thompson, 1987; "quantum mechanics", 2016). In addition, the concept used by microwave and laser were explained including other commercial goods that were made such as micro wave stoves, electron microscopes, and laser pointers.

Many well-known physicists supporting quantum theory to explain the tiny world of subatomic level received many Nobel prizes. For example, Sir James Chadwick (1891-1974), the winner of the Nobel Prize for Physics in 1935, discovered neutral particle called neutron ("James Chadwick - New World Encyclopedia", 2016). Wolfgang Pauli (1900-1958), winner of the Nobel Prize for Physics in 1945, used his principle called "Pauli Exclusion Principle" with quantum numbers to classify electrons of all elements ("Wolfgang Pauli - Biographical", 2016). Pieter Zeeman (1865 –1943), winner of the Nobel Prize for Physics in 1902, explained that light has its poles ("Pieter Zeeman", 2016). Enrico Fermi (1901 –1954), winner of the Nobel Prize for Physics in 1938, explained that uranium decay made new chemical element. After that in 1938, Otto Hahn and Fritz Strassman detected barium after bombarding uranium with neutrons ("Otto Hahn, Lise Meitner, and Fritz Strassmann | Chemical Heritage Foundation", 2016). The Fermi's work with Beta decay explained there were the relation among neutron, proton, and electron ("Enrico Fermi - Facts", 2016). Right now, physicists accepted that Fermi's theory was the first theory of the Weak Interaction that leads to one of four fundamental forces in universe (Kaku and Thompson, 1987; Parker, 1987).

Cyclotron, the particle-accelerating machine, was used to study the detail in an atom of elements ("The First Cyclotrons - Ernest Lawrence and the Cyclotron: AIP History Center Web Exhibit", 2016). Many physicists who concerned it were Robert Van de Graaff (1901-1967), who built the Van de Graaff generator, the first particle accelerator, to demonstrate the effects of high DC potentials ("Van der Graaff generator", 2016). John Cockcroft (1897 –1967) and Ernest Walton (1903 –1995) who worked in 1932 on the acceleration of protons bombarded lithium with high energy neutrons, electrons and protons and succeeded in transmuting it into helium and other chemical elements. This was one of the earliest experiments that changed the atomic nucleus of one element to a different nucleus by artificial means. This event was known as splitting the atom nucleus. Their work was instrumental in the development of nuclear power ("John Cockcroft - Facts", 2016; "Ernest Walton - Dictionary definition of Ernest Walton | Encyclopedia.com: FREE online dictionary", 2016; "John Cockcroft", 2016.). Otherwise, Ernest Lawrence (1901 –1958), winner of the Nobel Prize for Physics in 1939 for his invention of the cyclotron, is widely known as a pioneering American nuclear scientist (Kaku and Thompson, 1987).

With cyclotron, physicists found and classified many particles in many kinds of groups. For example, Hadrons consists of Meson and Baryon such as proton and neutron. On the other hand, Leptons is considered to be the fundamental particles, where physicists use the weight of the particles as the criteria of particle classification. Moreover, physicists use the spin of particle as the criteria for dividing the fundamental particles into two groups: Bosons and Fermions ("Hadrons - DiracDelta Science & Engineering Encyclopedia", 2016). Notably, Bosons are the fundamental particles which were named by Indian physicists, Satyendra Nath Bose (1894 –1974), which include Proton, W-particle, Graviton, and Pi Meson. In addition, Fermions are the fundamental particles such as electron, neutron, proton, neutrino, and quark ("boson | subatomic particle", 2016).

A brief of Einstein's work

Einstein was the first physicist of the world who gave us new characteristic of time- time as a relative value, not an absolute value. The big change of space-time knowledge was explained by Albert Einstein that I call Einstein's space-time in 1905. Before that, in 1865, at Cambridge University, James Clerk Maxwell, one of the brilliant physicists of the world, integrated the magnetic force and the electrical force into one coherent picture called an electromagnetic force (Kaku and Thompson, 1987; "James Clerk Maxwell - New World Encyclopedia", 2016). At that time, it was incredible that the magnetic force that makes a compass point to the north and the south, and the electrical force that makes our hair on our head stand when amber bar after rubbing with leather was brought near our head, would be united. But with his theory, right now, we are using dynamos to make electrical current for all electrical devices in our home because the electric field can create the magnetic field and vice versa.

The equation of Maxwell predicted that two sources of electromagnetic waves were able to be interfered. Moreover, the electromagnetic wave was the constant velocity, and looked like the water wave with very high speed. Light, ultraviolet, infrared, gamma ray, and x-ray are the other electromagnetic waves that exhibit the speed of light. The different frequency of wavelengths was used to create a distinct classification since each wavelength was unique, with the long wavelength at the top of the wave and longer than the shorter. The movement of the electromagnetic wave looked like when we play dominos by putting the white ones and the black ones repeatedly, and after we pushed the first domino down the others would be simultaneously pushed.

Overview of the second Einstein's Relativity theory

In normal life, if we are driving on a highway, say from the north to the south, at 60 miles per hour; the measurement of velocity of the car moving from the south to the north at 50 miles per hour, would give 110 (60+50) miles per hour. In other words, we get the accumulation of the speed of two cars when they are driven in the opposite directions by simply adding the two velocities. In contrast, if we are trying to measure speed of a car running from the north to the south at 70 miles per hour, the result is 10 (70-60) miles per hour. That implies that the total speed of the two cars comes from the subtraction of their speeds. Otherwise, the total velocity depends on the direction and the speed of two cars.

In terms of the speed of light, the big difference from the previous measurement is based on various aspects including its unique features. Light has a special characteristic; whereby, a person moving in the opposite or same direction as light will obtain the same speed. In other words, although we might measure the velocity of light in the different directions (the horizontal and the vertical directions comparing the earth), the speed of light will always be constant. As a result, Albert Michelson and Edward Morley got the Nobel Prize in physics because of this finding.

In addition, Maxwell's electromagnetic theory predicted that the special condition could make time shorter (Kaku &

Thompson, 1997; "Electromagnetic theory", 2016). This point was the beginning of the first Einstein's theory called the Special Theory of Relativity. At this time, Einstein investigated the characteristic of time in the Maxwell's theory before presenting his first theory in 1905. Following the Einstein's Special Relativity Theory, the reason why the velocity light measured as a constant value is based on the time. Specifically, an object moving at the light's speed is slower than the time when the same object moves at low speed. Otherwise, time is considered to be a relative value, where it depends on frames of reference. Besides, anybody can measure the different values of time at different places in the universe.

The conservative physicists did not accept the first Einstein theory until 1909 until Henri Poincare (1854-1921), the well-known scientist in that time, presented his work, which was similar to that of Einstein (Kaku and Thompson, 1987; "Henri Poincaré - New World Encyclopedia", 2016). However, the Einstein's General Relativity theory is different from the Special Relativity theory (Eisinger, 2016). The theory explains that huge gravitational force of a star that the light passes bends the light in space. For many years, with the help of his friends, mathematician Einstein solved mathematically to prove his second relativity theory. When the second Einstein's theory was presented, many people including physicists that supported the quantum theory said that the prediction of the theory was impossible. Before the day of scientific proof, some students that attended Einstein's class asked him what he was going to do in the event his prediction of the theory was wrong. He confidently said he could express his apologies to God; otherwise, he was confident that his theory was right.

The opportunity to test the Einstein's General Relativity theory was in May 29, 1919, when the solar eclipse occurred. Sir Arthur Eddington and his colleagues pointed out that if light was bent because of gravity force, then he and his team standing at different places of the earth would take some photos of a star behind the sun that could not be seen in a normal situation (Kaku and Thompson, 1987; "Arthur Eddington facts, information, pictures | Encyclopedia.com articles about Arthur Eddington", 2016).

Sir Eddington and his team photographed the target star while solar eclipse was occurring. The first point of their pictures was done at the Principe Island outside the shore of Western Africa, while the second one was taken at the Sobral city in Brazil. The results showed that they team was able to observe the star by two points instead of only one point. It was this time when the reporters commented against Einstein, with the photos providing the concrete evidence to complete the achievement of Einstein's theory.

The review of the development of the Einstein's theory established some remarkable views. Notably, it was found that

Einstein questioned himself about what light looked like in case where the light is viewed under the Newton's law of gravitational force. The solution was divided into two steps. First, how a woman or a man would be when s/he was under the acceleration. An example in this scenario is provided in *figure 1*, where a woman in an elevator gets an accident and falls down from the height without any slings.

In this situation, following the second Newton's law, the woman was falling under the constant accelerating force. Following the third Newton's law, the force on earth that pulls down equals to the force on the floor that push her feet up; thus, the force at the feet would become zero and the woman would float in the elevator in condition of weightlessness.

Notably, upon breakage of the sling, the emergency light at the top left corner of the woman's head in the elevator would be automatically opened. Consequently, light would move from the left to the right sides of the elevator's wall and the woman would be under the Newton's law of gravitational force.

Figure 1: Light running under gravity following Einstein's General Relativity Theory

Figure 1 above shows how light moves in the curve shape, as described by the second relativity theory of Einstein.

At time, t = 0, the sling had just broken when the elevator was at the top of the building. Consequently, the light began to move from the left to the right. After that, at t = 1, the elevator fell to another position far from the top of the building, while the light travelled from the left to the right sides. Hence, people who were observing the situations would see the light travelling in a shape of a curve, but the woman in the elevator would see the light travelling in straightforward direction.

Some of the leading scientists have also conducted various experiments to test the Einstein's General Relativity theory. One of them, Robert Pound, a Professor of the Harvard University in 1959 and 1965, shot the Gamma ray from the top of the building 74 feet height to the ground. The result of the test revealed that the wavelength of the gamma ray was significantly changed. Overall, the experiments supported the second Einstein's theory again (Hoffman, 2014).

A brief of Minkowski's work about time

Hermann Minkowski, an Einstein's teacher of mathematics at the university, explained the theory of Einstein's Special Relativity with his theory known as the Minkowski's 4D space-time ("Hermann Minkowski Facts, information, pictures | Encyclopedia.com articles about Hermann Minkowski", 2016). The time dimension was used to describe the mathematical model as (x, y, z, t). Minkowski indicated that the time dimension in a three-dimensional body located among three principal axes called "Path of Time Dimension."

See Figure 2.

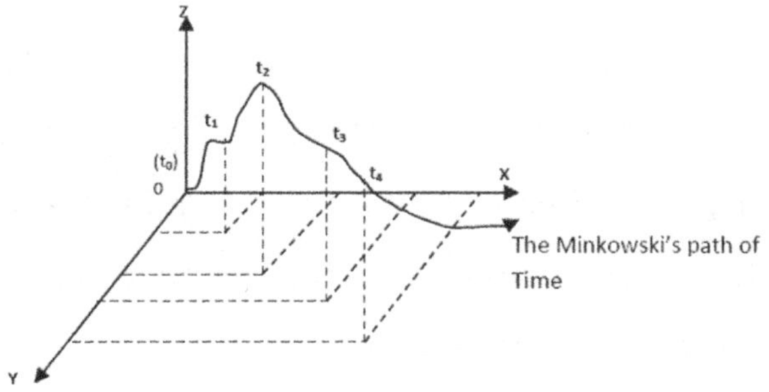

Figure 2: The path of the time dimension following Minkowski's 4D space-time.

Figure 2 shows how time locates among three principal axes. In the figure, time is separated from x, y, and z following

Minkowski's model. In other chapters, the Minkowski's model is used to make conclusions that it is not right because time values are in all Xs, all Ys, and all Zs; thus, it is not possible to separate time from x, y, and, z.

A brief of Hawking's work about time

Stephen Hawking, the Lucusian Professor of the University of Cambridge, the author of a popular book "A Brief History of Time," pointed out some parts of the General Relativity theory that were not conclusive. Professor Lucusian supported his view by the occurrences of the Big Bang; whereby, prior to its occurrence, the universe was not in the state of Singularity with the bend of space-time being the infinite value.

Further, Hawking explained his idea about the time by using a number system and the concept of Minkowski about the time dimension in the Light Cone. Presently, physicists and students of physics especially time, have to study Minkowski's 4-D models as one of the crucial concepts about time. In Minkowski's 4-D model, it is possible to highlight both the past and future aspects of time. In the past, Light Cone considered time to be the negative values; hence, such movements to the past would be possible for us.

The Minkowski's 4-D models have been used in many cases. Notably, many filmmakers used Minkowski's 4-D model to support their movies. Also, Hawking used Minkowski's 4-D model to support the second chapter of his book, which is one of the bestseller books as well. Physicists who did their researches on time and space, used the mathematical model of Hermann Minkowski to explain their results ("Hermann Minkowski Facts, information, pictures | Encyclopedia.com articles about Hermann Minkowski", 2016).

The numbers that are used in the model can be divided into two; the real numbers, which consist of the rational and irrational numbers, and the imaginary number. The real numbers can be plotted in the Light Cone; however, the imaginary numbers cannot perform the difference between the time and the space in the Light Cone (Hawking, 1998). Further, Hawking used the phrase "arrows of time" to explain time in three different kinds: Thermodynamics, psychology, and the expansion of the universe. For example, when the time passes, the energy from foods used to perform various activities would decay. Also, when the expansion of the universe occurs, lack of the discipline of objects in the sky would increase. Nevertheless, the arrow of time is not the direction of time dimension. Overall, the time dimension in the view of present physicists is considered to be the relative value but does not have a unique direction; thus, its scalar values.

The difference of time in quantum mechanics and Einstein's theories

A comparison between the paradigm of quantum and Einstein's relativity theories is shown in the table below.

Table 1: Comparison between two paradigm of physics

Items	Quantum mechanics theory	Einstein's Relativity Theory
What for?	To solve the problem how subatomic level looks like, how many kinds of particles are there in an atom, what are characteristics of particles?	To solve the problem how light looks like at the highest speed in space,
creators	Dirac, Heisenberg, and Schrödinger,	Einstein
What world?	Explains the world of the atom,	Explains the part of the world of celestial bodies,
About data?	Some data were used first, and equation from experiments	Maxwell used some data of Faraday, and his equation from his thinking
About time	No new knowledge of time dimension presented.	Time dimension was a relative value, not absolute value.
Clue in Maxwell's equation	Three physicists did not see little clue about time. Dirac had found Schrödinger's equation did not work well with Einstein's Special Relativity Theory before he created his new equation.	Einstein saw little clue about time and speed of light; time as the relative value; speed of light measured by everyone with the same speed.

Mass and energy	Nuclear fission of heavy elements was discovered in December 1938 by Otto Hahn and Fritz Strassmann, and explained theoretically in January 1939 by Lise Meitner and Otto Robert Frisch; this was used to explain that it was possible to make an atomic bomb.	Einstein had explained the relation between a little mass and enormous energy with $E=mc^2$ in 1905 before physicists found nuclear fission in 1938.
Develop ment	The theory still has been developing for hundred years (until now) from be simple to complicated models.	The new characteristics of time were not found after Einstein's Special Relativity had been presented.
Accepted by physicists	The new discovery in quantum such as nuclear fission was tested out hundreds of times in the course of 30 years.	When the first and the second Einstein's Relativity Theories were presented, no physicists studying quantum theory agreed with.

In the table, the paradigm of quantum theory originated from three leading physicists, Schrödinger, Heisenberg, and Dirac, who explained the world at the subatomic level. The development of the theory came from Max Plank who used some data from experiments. Interestingly, the three leading physicists neither gave new characteristics of time nor saw any clue about time before they passed away. However, Dirac only found Schrödinger's equation not work well with Einstein's Special Relativity Theory and eventually created his new equation. Thus, it is Dirac's equation that became the second step of two great theories.

Physicists who studied quantum theory used fission reaction to explain how a little heavy element changes to new lighter elements with emission of enormous energy. Nuclear fission of heavy elements was discovered on December 17, 1938 by German Otto Hahn and his assistant Fritz Strassmann. Further, in January 1939, Lise Meitner and her nephew Otto Robert Frischand theoretically explained that it was possible to make an atomic bomb. The knowledge on an atom was developed from two particles, electron and proton to many particles. In other words, quantum theory was developed from simple to complicated models. When new discoveries in quantum theory were presented, physicists tested and finally accepted it, but Einstein's discovery provided a different paradigm with quantum theory. These were not tested by physicists studying quantum theory ("Otto Hahn, Lise Meitner, and Fritz Strassmann | Chemical Heritage Foundation", 2016).

The beginning of the first Einstein's theory came from Maxwell's explanation of the magnetic and electrical forces working together as one coherent picture by the use of some data from Faraday's experiments. Einstein's theory gave new characteristics about time dimension, where time was described as a relative value and not an absolute value. In addition, Einstein explained the relation between a little mass to enormous energy with the equation $E = mc^2$ in 1905 before physicists found nuclear fission in 1938. Einstein's thinking has been widely adopted in the later physics compared to the quantum theory for more than thirty years. Unfortunately, after Einstein had presented his first theory, the new characteristic of time was not found again. The paradigm of second Einstein's Theory came from Einstein's trying to solve the problem how light looked like when we move at the speed of light. This paradigm is used to explain the part of the world of celestial bodies. When the first and the second Einstein's Relativity Theories were presented, no physicists studying quantum theory agreed with; those papers were refuted.

Whereas Quantum theory is used to explain the world of an atom; Einstein's theory explains the part of the world which comprises celestial bodies. It looks like pieces of a whole-unique picture of nature were placed in two different parts. However, nature does not depend on any theories by physicists. In contrast, scientists try to understand the whole nature. Besides, theories are created to explain some parts of nature as a whole. Because of the findings on new characteristics of time, science and social science may be changed in the near future.

The two great theories came from different paradigms; which include the paradigm of time dimension and the paradigm of quantum theory. Unfortunately, the paradigm about time was not important just like the paradigm of quantum theory in the context and view of the physicist studying quantum theory. Notably, Einstein has never received any Nobel Prize in physics from his special and general relativity theories that concern the paradigm about time. He only got the Nobel Prize award from his work that supported the paradigm of quantum theory, which is known as the Photoelectric Effect. This indicated that most of present scientists who got Nobel prizes were interested in quantum theory and might be one of the crucial reasons why the knowledge about time was developed a little for a hundred year after Einstein had found the first characteristic of time in 1905.

The unsolved questions and changes in physics

Right now, despite the fact that, these first-rate theories are running properly of their personal one of kind worlds; however quantum could not explain the celestial world. In addition, Einstein's concept couldn't run nicely in the subatomic stage. They could not hyperlink collectively. This shows that there may be something "incomplete" in both of them.

There are numerous unsolved questions in physics. For instance, inside the tiny world of the subatomic level, as an instance, the quantum theory itself could not provide an explanation for the motive why the widespread magnetic pressure that kilos matters into dust is launched while an atomic bomb explodes. Similarly, physicists honestly recognize the lifestyles of the magic perspective that suggests special function; however they do no longer know why it exists inside the subatomic stage. Physicists reading quantum paradigm cannot solution their own most basic questions: wherein is the part of the subatomic stage generating the magnetic pressure? And why is there the magic angle inside the subatomic degree as properly? (Bydder et al., 2007). In celestial global, the black hollow continues to be wondered many instances without answers ("hubblesite: black holes: gravity's relentless pull interactive: encyclopedia", 2005).

In the past, we saw big changes in our societies when new technology and innovation took place. The invention of light bulbs, the phonograph, the steamboat, the computer, and i-pad changed the structure of society and economic consequences. The Walt Disney Company was not the same when the company bought Pixar Animation Studios in 2006 at a valuation of $7.4 billion. Future changed because of new paradigms on innovation ("Disney buys Pixar - Jan. 25, 2006", 2016).

Some physicists expected that physics would reach to the advanced level of technology that no one knows if “Theory of Everything” is found. All equations in physics will be explained by an equation. Some of scientists also believe that in the future, the new knowledge of Theory of Everything or Unified Field Theory may be a great step to open the new era of the extreme technology on aviation that we have never seen before. Will we witness to the genesis of the theory of everything or Unified Field Theory in the future?

Hermann Minkowski explained that the time dimension in his 4-D model of space-time is separated from three principal axes with the form of Cartesian coordinate as (x, y, z, and t). There is no time value in the three principle axes. The three principal axes following the model without time inside them link to time traveling, one of the dreams of filmmakers. Minkowski thought that numbers with positive and negative values having positive and negative time values respectively. In other words, Minkowski's 4-D model pointed out that each positive number had positive time value; and negative numbers had negative time values. If time values are negative in past Light Cone, then time traveling can be possible.

Filmmakers who believed in the Minkowski's 4-D model of physicists have made many movies, such as "Back to the future," and "X-Men: Days of Future Past" for entertainment. The performers in the movies wanted to change something "wrong" in present; they went back adventurously to the past for changing some crucial events. When some people asked filmmakers about possibility of the stories, the makers always said that the stories based on physics' theory. Is it possible for us to travel to past for some mission?

A brief crucial time line of the first characteristic of the time dimension

Aristotle trusted that the time measurement constantly connected to space, called space-time. Both time and space were total qualities, which did not rely on upon any edge of reference. As such, the time and the space were the uncommon qualities since they were measured in similar values in all spots of the universe ("Time | Internet Encyclopedia of Philosophy", 2016).

The Aristotle's space-time may be resolved from 300 B.C to 1600s; where Aristotle envisioned that the time could be spoken to as a flat plane, and the space as the vertical plane (normally called world line). Galileo's space-time may be resolved from 1600s to 1905. With the Galileo's Principle of Relativity, time and space which were clarified uniquely in contrast to Aristotle. Galileo trusted that the time was still the supreme esteem yet the space was the relative esteem. He envisioned that the time measurement was a "heap of planes" which was inclined ("deck of cards"). Newton put stock in an indistinguishable path from Galileo too (Norton, 1999; "Galileo Galilei - New World Encyclopedia", 2016; Wikipedia, 2015).

In 1865, the Maxwell's electromagnetic theory predicted that there was the special condition which made time shorter ("Electromagnetic theory", 2016; Kaku & Thompson, 1997).

In 1887, Albert Michelson and Edward Morley measured the speed of light and concluded that the speed of light was always constant in the different directions of measurement ("Albert A. Michelson - Biographical", 2016; "Edward Morley - New World Encyclopedia", 2016).

In 1905, Einstein explained why the light speed is constant in his Special Relativity Theory that the time was a relative value. In other words, the time dimension depends on a frame of reference. When we are in a rocket traveling at very high speed, the time in the rocket was slower than the time at the ground of the earth (Kaku and Thompson, 1987; "Albert Einstein | German-American physicist", 2016).

Some scientists explained the theory to twins who are traveling at the different speed in the different frames of reference. While the first one was rocketing in space at a speed of nearly 300,000 kilometers per second or 186,000 miles per second, the speed of light, and the second one was living on the earth at a speed of a point near the equator moving at close to 1000 miles per hour (1600 km/hr.) ("ASP: How Fast Are You Moving When You Are Sitting Still?", 2016). The frame of reference of the first one was a rocket, and that of the second one was the ground. Ten years later, the twins came to meet each other;

the first one was younger than the second one because the time in his rocket was slower than that on the earth. In other words, the time dimension is a relative value, which measured the difference at different frames of reference.

In addition, the theory also points at the speed of light as the highest velocity in the universe, no one moves with the speed of light because going faster than the speed of light requires infinite energy.

Finally, Einstein explained the relation between energy and mass by using the formula, $E = mc^2$. Where E is the energy of a physical system, m is the mass of the system, and c is the speed of light in a vacuum (about 3×10^8m/s). In other words, energy equals mass multiplied by the speed of light squared ("How "Fast" is the Speed of Light?", 2016).

The conservative physicists accepted the first Einstein's theory in 1909 after Henri Poincare (1854-1921), the leading scientist in that time, had presented his work which was similar to that of Einstein (Parker, 1987; "Jules Henri Poincare facts, information, pictures | Encyclopedia.com articles about Jules Henri Poincare", 2016).

In 1988, Stephen Hawking explained his idea about something new called "the arrow of time" was presented for explanation of only the trend of things; it is not a real direction of time (Hawking, 1988).

In 1995, Paul Davies wrote his book "About Time." He indicated that many views of the time dimension which is a missing link that scientists have not known clearly yet (Davies, 1995).

In short, the time dimension is the relative value, and it has only its magnitude not its direction.

The time dimension is in everything in the universe

Time is a unique missing linkage that can connect the universes of science and sociology together. The ruler of science world is material science and the ruler of sociology world is financial matters. Time is concealing itself in all measurements in both universes. In the realm of science, for example, material science, cosmology, science, science, engineering, pharmaceutical, software engineering, geography, meteorology, human studies and else, time is on the whole. In the realm of sociology, for example, financial matters, promoting, human asset administration, association, coordination and store network, tourism, law, religion, history, phonetics, music, convention, custom, verse, expressions, soothsaying and else, time likewise is in them.

In addition, time links all dimensions together. For example, the huge things are mountains, planets, stars, the sun, galaxies, and the universe. All the huge and the small things have their sizes. In the past, people believed that an atom is the smallest thing in the universe. Right now, with quantum mechanics theory, most physicists found many particles such as an electron, a neutron, a proton inside subatomic level. All things are in the three-dimensional world that consists of a length, a width, and a thickness.

Moreover, time links all activities in economy together. Production, exchanging, distribution, and consumption, which are the main four economy activities, are linked together by the times of producers, exchangers, distributers, and consumers. In real world economy, there are many dimensions in a price and time values are in all. For example, if the buyer H pays $50 for a shirt, then the price of the shirt comprises many prices which are made by many sellers in the different time values. The producer E sells a shirt 20 dollars each to a wholesaler F. In this case, the producer pays average four dollars each for a piece of cloth of an owner A, average one dollar each for his or her worker B, average two dollars each for the landlord C, average three dollars each for the banker D, and average 10 dollars each for producer's profit. The time values are in all dimensions, and the time dimension is linking all dimensions together.

The wholesaler F and the retailer G make the second and the third prices 30 and 50 dollars each for the goods in different time values respectively. Both of them make their prices for running their business.

Time links all places together as well. If we want to visit our parents at their houses in different states, we have to lose our time for traveling. The time values are in everything that has been changing all the time.

Everything is also under control of time. Everything has its own time as well. For examples, we have to spend a year for growing a banana tree to produce its product. Similarly, we use five years for growing a mango tree to produce the first mango. Mr. B was born on May 26, 1940, and passed away on February 6, 1999. While the summer is in the northern region of the earth, the winter is in the southern region. When New York is at night, Bangkok that is the capital of Thailand is at day simultaneously.

Numbers representing everything

Everything in the universe has two main kinds: tangible and intangible dimensions. For example, the tangible dimensions are the things we can touch such as all kinds of matter, cells, organs of human being, animals, grasses, leaves, trees, rocks, mountains, and planets, stars, galaxies, and the universe.

Mathematicians use numbers to represent everything in the universe. Human created numbers for counting things in the past. The numbers were used for counting their animals to protect some lost, to count their food in barter system and to count days for cultivating. We used the numbers to represent things that we can see, hear, smell, taste, touch, including things in the ground, under the sea, in the sky.

The intangible dimensions are some of things we cannot touch. However, they can be measured or counted using electronic devices and other methods such as force, energy, and wave, particles, utility, pleasure of human being, emotions like fear and love. Time in mass, energy, force, and field can also be found. For example, time can be found in the unit of force in an acceleration, which is the change of velocity by the change of a time value (or $a = \frac{\Delta v}{\Delta t}$) in Newton's equation, $F = ma$. Also, time is in magnetic force equation: $F = q\,(E + v \times B)$ when v is the particle's velocity ("magnetic force | physics", 2016). Time is in fields such as electromagnetic field, the field (as the magnetic induction equation, $\frac{\partial \mathbf{B}}{\partial t} = \eta \nabla^2 \mathbf{B} + \nabla \times (\mathbf{u} \times \mathbf{B})$) where u is the velocity of the fluid ("electromagnetic field | physics", 2015). Time is in energy because there is time in "c" of Einstein's equation, $E = mc^2$. Time is in kinetics and potential energy. Time is in mass because "F = ma" means m = F/a. These are the evidences that the time dimension is in forces, fields, energy, and mass. Scientists in all fields always use the numbers for their quantitative researches.

In addition, time still is a crucial dimension that most economists have overlooked for a long time. In the Figure 3.a and 3.b, the time values were found in the price neoclassical theory. See Figure 3. The adjustment of prices to the equilibrium point, t = 1 and t = 3, are the time values at the beginning of the adjustment. In addition, t = 2 and t = 4 are the time values which the adjustment completes respectively.

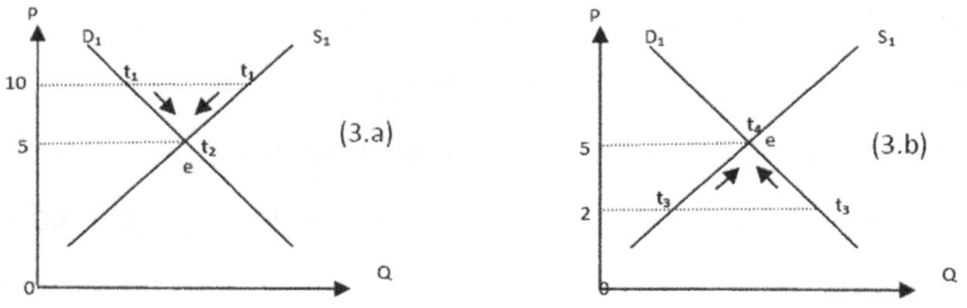

Figure 3: Time in the adjustment of prices to the equilibrium points

Figure 3 shows that there is time in the equilibrium theory of microeconomics. The theory indicates that if a price occurs over or under the equilibrium price, then the price will either decrease or increase respectively up to the equilibrium point. Time is hiding itself in two figures.

In addition, the change of Marshall's equilibrium theory in comparative statics needs different time values from the beginning to the end ("John Richard Hicks - New World Encyclopedia", 2016.). See Figure 4.

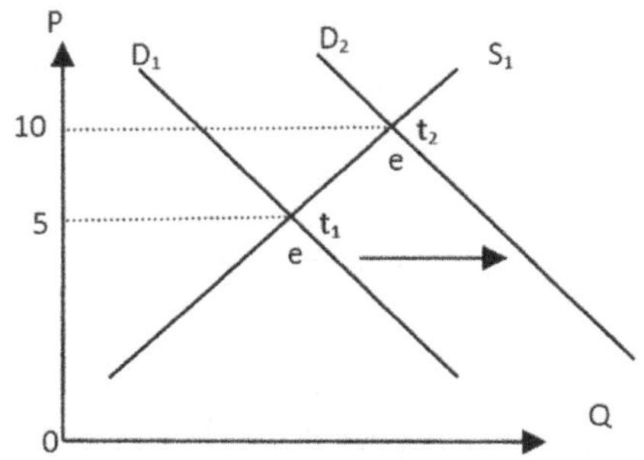

Figure 4: The shift of the Marshall's demand curve from D1 to D2 has to spend time.

Figure 4 shows the presence of time in the price theory of microeconomics. The figure shows where the different time values occur. The theory indicates that if an increase in demand such as a rise in income or a change in tastes that favors the product, then the demand curve of the buyer will shift from initial position to the right hand side. It is important to appreciate the fact that the curve moves from the initial position to the new one; thus, this activity needs to have different times. Besides, the theory does not emphasize about time, it simply shows how the two curves are used in the comparative statistics.

Moreover, the price adjustment to equilibrium of Cob Web Theory needs to have different time values in all steps of changing as well (Pashigian, 2008; "Cobweb model", 2016.). See Figure 5.

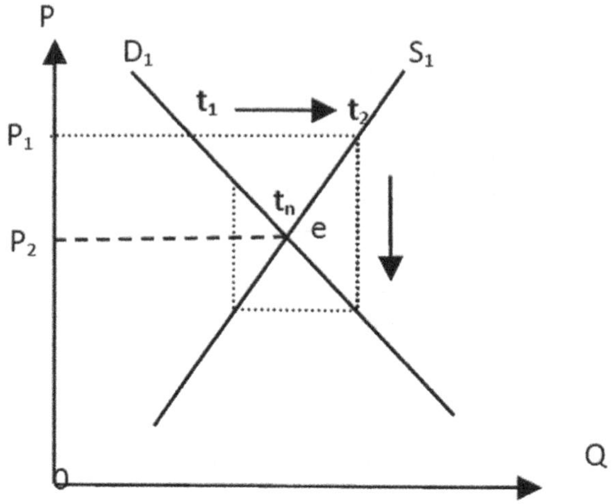

Figure 5: In Cob Web Theory, the adjustment of a price to the equilibrium point has to spend time.

Figure 5 shows that there is time in the Cob Web theory of microeconomics. According to the theory, price adjusts itself to the equilibrium price or out of the equilibrium point. That implies that when there is adjustment moves from the first position to the new ones, then these activities need to have different times. The emphasis in the theory is not but the directions of adjustment, where arrows are used around the equilibrium point called Cob Web theory. In addition, time dimension is the center of problems in economic theory.

In the past, Marshall, the economist who created the concept of the demand and the supply curves in microeconomics, commented upon time that "The element of time is the center of the chief difficulty of almost every economic problem" (Alfred Marshall quoted in Davidson, 1981).

In 1985, Robert Kuttner wrote an article "The Poverty of Economics" in Atlantic Monthly to indicate that something wrong is still at the core of economic theory. If economic theories are wrong, then policy-makers who have economists as their economy consultants will make wrong policies and wrong directions to solve economy problems. The wealth of the nation in short and long run is too far for all people of the country to reach that dream. The weak point in economic theories is not a huge problem in only its field, but it can make economy problems at the world level. For example, if policy-makers of China, Russia, Japan, and the USA agree with a plan for solving problems about greenhouse effect phenomenon, then in the near future, some activities of solving all problems concerning the effect will be set.

In the end of the article, Kuttner referred to the comment of 20th-century economic historian Robert Heilbroner. For example, the economic theory could not answer many basic questions such as what a price is, what money is, what killed full employment, what the boundary between economics and politics, or society is ("The Poverty of Economics" by Robert Kuttner, The Atlantic Monthly, February 1985", 2016). (The time dimension concerning economics was discussed in the seventh chapter.)

Is the time dimension in the three-dimensional body especially in a subatomic level?

The smallest world is in subatomic level. In subatomic level, after the well-known physicists such as Cockcroft, Walton, and Lawrence had developed their cyclotrons, the particle accelerating machine, physicists found and classified a lot of particles in many kinds of groups.

Hadrons consisting of Meson and Baryon such as proton and neutron, and Leptons are the fundamental particles that physicists use the weight of the particles for particle classification. Moreover, physicists use the spin of a particle for dividing the fundamental particles into two groups; Bosons, and Fermions as well ("boson | subatomic particle", 2016).

Bosons are the fundamental particles, which are named after Indian physicists, Satyenda Bose, such as Proton, W-particle, Graviton, and Pi Meson. In addition, Fermions are the fundamental particles such as electron, neutron, proton, neutrino, and quark ("Boson", 2016).

If there is time in everywhere in the universe, so the time can be found in subatomic level as well. Right now, there is an angle called by scientists “magic angle” in subatomic level. It has its own special characteristic as well. (Charpentier, Fermon, & Virlet, 1998; Bydder et al., 2007)

The magic angle, which is approximately 54.7356°, is defined in the field of nuclear magnetic resonance spectroscopy, and in the field of electron energy-loss spectroscopy. At the magic angle, some interaction in subatomic level vanishes; as a result, scientists are using it for physicians to heal their clients more efficiency. For example, scientists studied magic angle spinning that is a technique in solid-state NMR spectroscopy which employs this principle to remove or reduce the influence of anisotropic interactions, thereby increasing spectral resolution. Finally, scientists invented one of scientific equipment to make magnetic resonance imaging. The image from this technique help physicians see more clearly structures with ordered collagen, such as tendons and ligaments that are oriented at the magic angle with appearing hyper intense in some sequences. Scientists call it "the magic angle artifact or effect" (Gaillard, 2016).

Scientists have known it for a long time, but they could not answer why the angle is different from the other angles. If the time can be found in this angle, we can conclude that there is the time dimension in subatomic level that is one of the three-dimensional bodies.

The relation among time, numbers, number lines, and three principle axes in a three-dimensional body

When we draw the line or the curve, our drawing has to spend our time. The more time we use the longer of the line or the curve we get. The faster we draw the line the longer of it we get as well. We concluded that the relation among distance, velocity, and time is real when we draw the line or the curve. In other words, there are at least three dimensions inside the line or the curve: distance, velocity, and time values inside the number line. In the line or the curve, there are a lot of continuous points; each point has its own value that is determined as each number. In other words, to show the different value of the point, mathematicians use a number. And each point occurs in the different time value. In other words, each point in the line or the curve has its own number and time value.

A number line has many numbers inside it. In a three-dimensional body, N number lines can be drawn to pass the origin. Although, there are N number lines inside the three-dimensional body but there are the only three important main lines called three principal axes to represent the length, the width, and the thickness. As a result, each principle axis is one of the number lines that are drawn to pass the origin. And each principal axis has a lot of numbers inside it. Each number inside a principle axis has its own time value as well.

If time is in everything, and everything is represented as numbers, so it should be time in all numbers. Numbers are in a number line; so time should be in the number line as well.

The hypothesis about where time is

The hypotheses about time were created; “is the time dimension in a number in a line?” and “is the time dimension in the magic angle?”

When we draw a line, two factors affecting the length are: (1) the velocity of drawing, and (2) the value of time (or time value) we use to draw. With the time value given, if a drawer uses higher velocity to draw, then the length of the line will be longer. Similarly, with constant velocity of drawing, if the value of time is longer, the length of the line will be longer as well. The Newton’s concept about the relation among distance (S), velocity (v = dS/dt), and time (t) is appropriate for this condition. In other words, the concept used to solve the hypothesis about a time value in a number in a number line, the positions of the time values at numbers in the line or the curve, is the Newton’s concept of the relation among distance S, velocity v, and time t; S = vt.

In this case, the origin as zero of the line is the absolute value and the frame of reference which all other numbers are comparing with. A point near the origin has smaller value than that far from the origin. The positive and the negative signs in front of numbers are showing the directions of the numbers. The both signs show the different directions as well.

Summary of the chapter

In 1865, the equation of Maxwell predicted that two sources of electromagnetic waves were able to be interfered, and, moreover, the electromagnetic wave was the constant velocity. It looked like the water wave with very high speed. In addition, an electromagnetic Maxwell's theory predicted that the special condition could make time shorter. After Michelson and Morley had measured the speed of light, they found that the speed of light was always constant in the different directions of measurement.

These two points were important parts of the beginning of the first Einstein's theory in 1905. Einstein explained that while we were moving with high speed, the time was slower than that with low speed. In other words, the time was a relative value, or the time depends on its frame of reference.

The conservative physicists accepted the first Einstein theory in 1909 after Henri Poincare (1854-1921) had presented his work which gave the same results of Einstein.

The travelling of twins that one in the frame of reference as rocket travelling in space with high speed and another in the earth was used for explaining the meaning of time as relative value. Ten years later, they met each other, as the result; the first one was younger than the second one because the time in his rocket was slower than the time on the earth. This was the big change of space-time knowledge explained by Albert

Einstein. Einstein's space-time might have been determined since 1905 until now.

Einstein changed the views of scientists about mass and energy after he had found the first characteristic of time dimension. Time is not a normal dimension that we know very well. In other words, we have just known about the details of the first time dimension from the first Einstein's theory. The knowledge of time is developed very slowly by present physicists who are studying quantum theory because the paradigm of time is in Einstein's theory. When the characteristic of time was presented in 1905, it is easily ignored by physicists at that time because physicists studying quantum theory know little about time.

The reason why Einstein's work was ignored is about paradigm shift. The paradigm shift is a situation a new paradigm that is rejected or ignored by people who could not catch the new paradigm. The new paradigm is always very different from the existing paradigm. People who hold to the existing paradigm will feel easier if they refute the new paradigm when they see the new paradigm for the first time.

When paradigm shift of something occurs, everyone has to be pushed to the starting point of the new paradigm following the law of paradigm shift. Experts become a new student in the class of new paradigm. This situation has taken place for many times in the history of physics. For example, when well-known physicists at that time saw Plank's equation, they would be pushed to the starting point of the Plank's work. They thought they knew everything important about physics; however, they knew nothing about Plank's discovery. They did not know they became students who had to learn the new paradigm in Plank's classes, not the experts. However, they felt the rejection of Planck's work was reasonable because they believed that their experiences were good enough to judge what was appropriate or inappropriate when they saw something "quite strange" for them.

In Switzerland, a Swiss inventor who invented the first quartz watch tried to demonstrate his invention to Swiss watch-making industry. Most of the members believed that they were the experts of the world of clocks and watches. When they saw the quartz watches, they did not realize that they were pushed to the starting point of a new-world clock. They judged the quart watches not to be good enough for "everyone." Finally, Japanese merchants of Seiko bought the patent of the new type of watches and became new watchmaking industry of the world. The Swiss watchmaking industry lost their market share and made Swiss employment between 1970 and 1988 fell from 90,000 to 28,000 ("Quartz crisis", 2016). The new Japanese watchmaking industry was born and became the first rank of the world of quartz watches. If Swiss watchmaking industrial realized that they knew nothing on new paradigm of future watches and clocks, they might keep their market share and still were the greatest country which produced mechanical and quartz clocks and watches in present.

Time is in everything in the universe. The new characteristics of time dimension may help people around the world to reach new understanding about nature of many things that we have never known them because of the paradigm of time making paradigm shift in physics.

The most important point in this chapter is as follow:

Time in everything + Numbers represent everything

Hypothesis 1: Are values of time in numbers?

Hypothesis 2: Are values of time in subatomic level?

Chapter 2
The Equation of All Equations

Intelligence is not the ability to store information, but to know where to find it.

Albert Einstein

In the first chapter, the hypothesis of the value of time in the same position of a number in a number line was created. In order to prove and indicate the exact position of the time value in the number in the number line, this chapter represents the Newton's equation, S = vt. The proof was divided into two steps; (1) the proof of existence of the time value in the line; (2) finding of the relation between a value of time and a number (called in this book "a determined number"), which is in a number line.

The time dimension in a number line: the first step

It is important to appreciate the fact that length, width, and thickness are the key subsets of distance. That implies every line has a definite distance. The starting point of a number line is known as its point of origin (or simply the origin). Two factors make up the distance from the origin to the end of the line longer are as follow: (1) the velocity of drawing, which we use to draw the line, and (2) the time value, which we lose for drawing the line. For example, given the time value as constant, the faster velocity of drawing a drawer makes, the longer distance of the number line we get. In addition, given the constant speed of drawing the line, the more time value we use, the longer distance of the number line we get as well.

Given the velocity of drawing as constant, when the time value we spend to draw the line increases, then the distance of the line is longer. The distance depends on the time value. The time value is the cause and the distance is the effect. In other words, the time value is an independent variable and the distance is a dependent variable or S = f (t). In this case, we can prove mathematically. See the proof in the part of Mathematical Proof of chapter 2.

$$S \quad = \quad f(t) \quad \ldots\ldots\ldots\ldots (1)$$

Equation 1: The distance is a function of the time value.

This equation forms the basis about the mysteries on time. The proof shows that all distances such as width, length, and thickness, which are the main parts of 3-D body, have time dimension in them. When we draw a line, there is time spent in such drawings. Various people imagine that the line depends on the speed of drawing (v) and the value of time; however the mathematical proof indicates that the line solely depends on value of time. Thus, any increase in the value of time, the distance becomes longer. Besides, the speed of drawing provides the ratio between two values of distance (with subtraction) and two values of time (with subtraction). Overall, this forms the origin of the relation between time and distance.

The distance in a 3-D body has three main sides: width, length, and thickness. Mathematicians use three principal axes (x-axis, y-axis, and z-axis) to represent the sides. They determine the positive values of numbers in the right or the upper or the front sides, including the negative values of numbers in the left or the bottom sides or the back. BUT all six sides have positive distance, which means that all side have positive values of time. Specifically, if the distance (result) is not negative, then all values of time (cause) cannot be negative too. This is an important aspect that the present scientists, including Hawking, believe that time in 3-D body is the fourth dimension following the model of Minkowski. In terms of numbers, x, y, and z do not have the value of time since time is separated by the new dimension in 3-D body. Otherwise, the mathematical proof shows that the line only depends on the value of time.

Other important aspects that can be appreciated from the mathematical proof are that the line has uncountable numbers with every number having its own values of time. In other words, time values are in all numbers. This proof shows that the model of Minkowski on time is an imaginary and not real. This proof opens minds of I and readers about time. It looks like something locked was opened. From this point, I acknowledged the new knowledge about time such as the time vectors, the time pyramid, the time in subatomic level, and the law of direction of time.

The function can be drawn as in figure 7. The function S can be drawn as the steep-straight line from the origin.

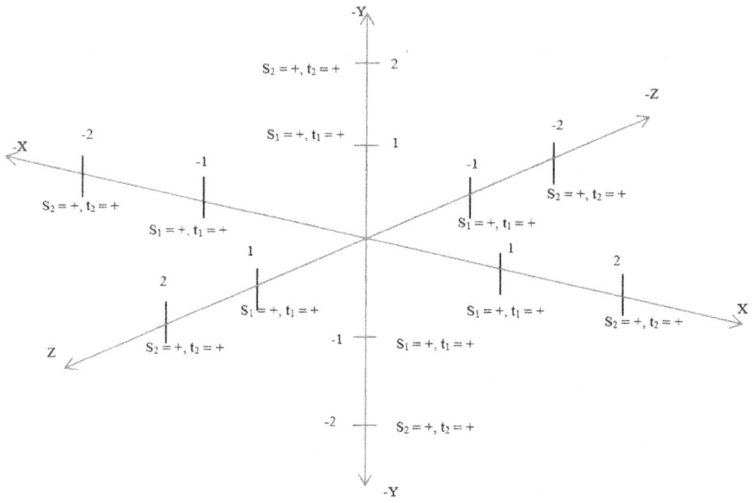

Figure 6: If a value of time increases, then distance increases

The figure shows the meaning of the first equation (S = v t). When a line is drawn, the values of time change from t = 1 to t > 1; thus, the distance of the line increases too in all directions. Otherwise, the numbers may be either positive or negative (+x, -x, +y, -y, +z, and –z), but the values of time and distance in all directions always remain positive.

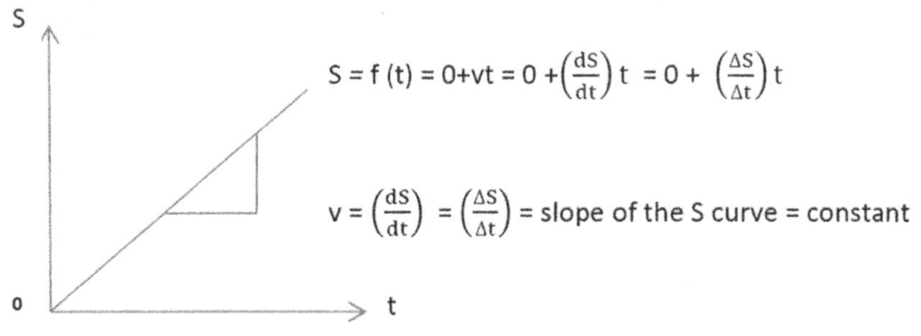

Figure 7: Distance is a function of the value of time.

In figure 7 the coefficient in the first equation is presented in terms of the distance as a function of the value of time. From the equation S = f (t), the coefficient of t is the slope of S curve. In this case, the coefficient of t is constant; thus, a straight line is obtained. Notably, the slope can be calculated by using two values on the vertical axis (S_2-S_1) then divided by two values on the horizontal axis with subtraction (t_2-t_1). The function becomes the scientific evidence of existence of time in distance. The second step is to solve what values of time and where in a line.

What a value of time and where in a number line: the second step

Different drawers have different ways to determine numbers in their lines. I called a number "a determined number." For example, three drawers can give different numbers at the same position of a line. See Figure 8.

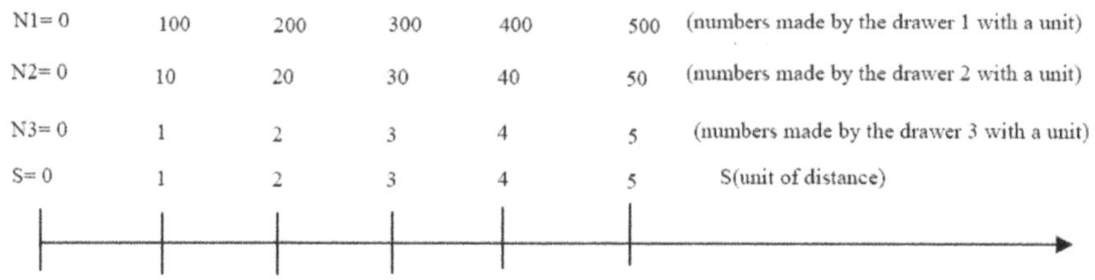

Figure 8: Different drawers may give different numbers at the same positions

The figure shows how "a position" at the number line can be "named" differently by different drawers (the drawers look like women giving different names of a baby). The other names will follow the first names.

For example, at S =1, the drawer No.1, 2, and 3 is giving the number as 100, 10, and 1 with different kinds of unit respectively. Similarly, at S =4, the drawer No.1, 2, and 3 is giving the number as 400, 40, and 4 respectively.

We can explain the relation between the determined number (N) in a number line and the distance (S) from the origin, except N and S as zero, as follow: $|N| = R.S$ when the symbol S represents the distance from the origin to the determined number. The symbol N represents the determined number. The symbol R represents the ratio of an absolute N by S. The symbol v or $\frac{\Delta S}{\Delta t}$ represents the velocity of drawing of the line. It is the percentage of the changing of distance by the percentage of changing of the velocity per unit of time. The symbol t represents the value of time which matches with the position of the determined number N.

For example, at the position from the origin (S) as two centimeters long, if the determined number is given -10 units; as a result, the value of R is five units per centimeter. In other words, the value of R is calculated by using an absolute value of ten by two, $\frac{|-10|}{2} = 5$. The distance S is always positive, so the value of R is always positive too.

We can write the relation among S, N, R, v, and t as follow:

$$\frac{|N|}{R} = S = vt = \left(\frac{dS}{dt}\right)t = \left(\frac{\Delta S}{\Delta t}\right)t \quad \ldots\ldots\ldots\ldots (2)$$

Equation 2: The relation among S, N, R, v, and t.

This equation is one of the most important aspects about time. Mathematicians determine the three pairs of numbers: (+x, -x), (+y, -y), (+z, -z); however, the relation between all numbers (with positive and negative) and distance cannot be negative. This is because the distance and the value of R always are positive. At this point an important aspect in mathematics is time, which is supported by the model of Minkowski; whereby, time can be negative in the Past Light Cone.

The value of time in a number line can be calculated as follow:

(1) the value of time in a line depends on the distance from the origin to the determined number and the velocity of drawing the line at the following equation;

$t = \frac{S}{v}$, (2) the time value also depends on the determined number with only the positive sign, the value of R, and the velocity of drawing at the following equation, $t = \frac{|N|}{R.v}$. The value of time is zero; it means that the drawer has not drawn the line yet. In addition, the determined number is zero; it means that it is a starting point of the line. These two formulas can be used to indicate the exact position of the value of time at the examples in following subtopic.

$$t = \frac{|N|}{R.v} = \frac{S}{v} = S\left(\frac{\Delta t}{\Delta S}\right) \quad \ldots\ldots\ldots\ldots (3)$$

Equation 3: The linear time equation

The equation is one of the most important aspects about time in various aspects. The relation among numbers (both positive and negative), distance, and the values of time can therefore be concluded. This point is a critical aspect about knowledge of time. The values of time can thus be calculated by; (1) using numbers (N), the value of R, and the velocity of drawing, and (2) using distance (S) and the velocity of drawing (v). The exact position of each value of time in all straight lines can then be indicated.

Normally, we always see the determined number but we have never seen the value of time matching with the position of the number (Boonsri, 2012; Boonsri, 2014). In the past, negative numbers were in use in India to represent debts; the numbers are the amount of money lent. In addition, Brahma –Sphita–Siddhanta, Indian mathematicians, used negative numbers to produce the general form of quadratic formula, which has been used until now. In the real world, the quantities of products have to be positive. For example, we say that we have ten boxes; in contrast, if someone says that he has minus ten boxes, this means nothing.

The exact position of a time value in one-dimensional figure

The relation among S, v, t, R, and N can be showed at the following figures. In Figure 9, if the velocity of drawing the line is two inches per second, and the distances from the origin to the five determined numbers are 20, 40, 60, 80, and 100 inches long. In addition, the first to the fifth determined numbers are 100, 200, 300, 400, and 500 respectively with unit as ton, then the values of time matching with the positions of these five determined numbers are 10, 20, 30, 40, and 50 seconds respectively. In this case, the R is five.

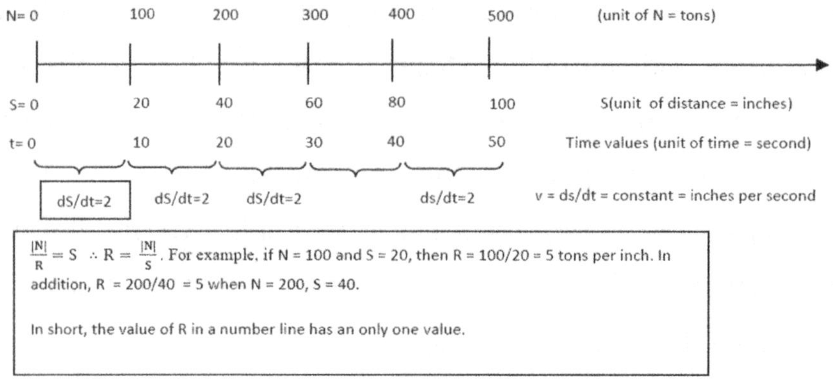

Figure 9: The relation among S, v, N, R, and t

The figure shows "a position" at the number line to have five values in it.

At N = 100 (tons), the distance S =20 (inches), the time value t = 10 (seconds), the line will be written by using speed as $(S_2-S_1)/(t_2-t_1)$ = (20-0)/ (10-0) = 2 (inches per second). In the event the number to the distance can be changed and vice versa, then the value of R ($R = \frac{|N|}{S}$) can be used.

For example, absolute of N = S.R or S = absolute of N / R.

(In this case 100 = 20. (5) or 20 = 100/5)

In *Figure 9*, we can calculate the value of time matching with the first determined number (100) as follow:

The first method by using S and v;

$$t_{at100} = \frac{S}{v} = \frac{20 \text{ inches}}{2 \text{ inches per second}} = \frac{20 \text{ inches}}{\frac{2 \text{ inches}}{1 \text{ second}}} = 10 \text{ seconds}$$

The second method by using N, R, and v;

$$R = \frac{|N|}{S \text{ inches}} = \frac{|100 \text{ tons}|}{20 \text{ inches}} = \frac{100 \text{ tons}}{20 \text{ inches}} = \frac{5 \text{ tons}}{1 \text{ inch}}$$

$$t_{at100} = \frac{|N|}{R.v} = \frac{|100 \text{ tons}|}{(5 \text{ tons per inch})(2 \text{ inches per second})} = \frac{100 \text{ tons}}{\left(\frac{5 \text{ tons}}{1 \text{ inch}}\right)\left(\frac{2 \text{ inches}}{1 \text{ second}}\right)} = 10 \text{ seconds}$$

In Figure 10, if the velocity of drawing the line is two inches per second, and the distances from the origin to the five determined numbers are 20, 40, 60, 80, and 100 inches long. In addition, the first to the fifth determined numbers are -100, -200, -300, -400 and -500 respectively with no units, then the values of time matching with the positions of these five determined numbers are 10, 20, 30, 40, and 50 seconds respectively. In this case, the R is five.

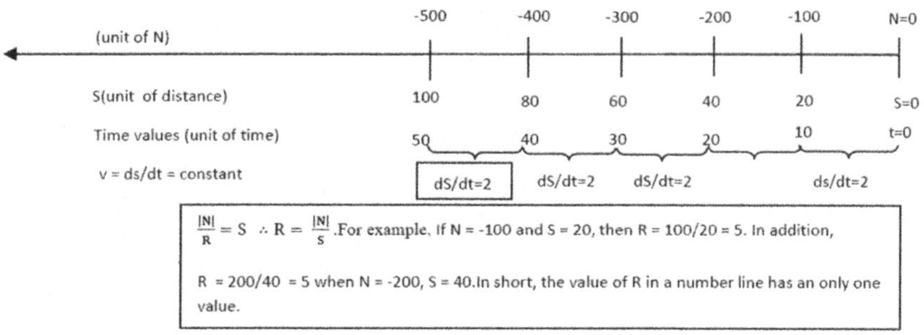

Figure 10: The time values are always positive, but the determined number may be negative

The figure shows "a position with the negative", whereby, at the number line, there are five values inside it. At N = -100, the distance S =20 (inches), the time value t = 10 (seconds). The line is written by using speed as $(S_2-S_1)/(t_2-t_1)$ = (20-0)/ (10-0) = 2 (inches per second). When there is need to change the number with the distance and vice versa, value of R ($R = \frac{|N|}{S}$) are used. For example, absolute of N = S.R or S = absolute of N / R. (In this case -100 = - [20. (5)] or

20 = [absolute of (-100)/5])

In Figure 10, we can calculate the value of time matching with the first determined number (-100) as follow:

The first method by using S and v;

$$t_{at-100} = \frac{S}{v} = \frac{20 \text{ inches}}{2 \text{ inches per second}} = \frac{20 \text{ inches}}{\frac{2 \text{ inches}}{1 \text{ second}}} = 10 \text{ seconds}$$

The second method by using N, R, and v;

$$R = \frac{|N|}{S \text{ inches}} = \frac{|-100|}{20 \text{ inches}} = \frac{100}{20 \text{ inches}} = \frac{5}{1 \text{ inch}}$$

$$t_{at-100} = \frac{|N|}{R.v} = \frac{|-100|}{(5 \text{ per inch})(2 \text{ inches per second})} = \frac{100}{\left(\frac{5}{1 \text{ inch}}\right)\left(\frac{2 \text{ inches}}{1 \text{ second}}\right)} = 10 \text{ seconds}$$

We can use the time equation (3) to find what values and where the exact positions of time in a line are. Every time value is matching with a position of a determined number. It looks like the time values hiding themselves at the same positions of the determined numbers in a number line. The equation (3) shows the linear relation between the distance and the value of time.

The positions of the time values plotted in three-dimensional figure

In Figure 11, if the velocity of drawing the line is 100 inches per second, and the distances from the origin to the three determined numbers are 50, 100, and 150 inches long respectively. In addition, the first to the third determined numbers are 10, 20, and 30 respectively with no units, then the time values matching with the positions of these three determined numbers are 0.5, 1.0, and 1.5 seconds respectively. In this case, the R is 0.02. Please notice that the values of time in this case are decimals

In Figure 11, we can calculate the value of time matching with the first determined number (10) as follow:

The first method by using S and v;

$$t_{at10} = \frac{S}{v} = \frac{50 \text{ inches}}{100 \text{ inches per second}} = \frac{0.5 \text{ inch}}{\frac{1 \text{ inch}}{1 \text{ second}}} = 0.5 \text{ second}$$

The second method by using N, R, and v;

$$R = \frac{|N|}{S \text{ inches}} = \frac{|10|}{50 \text{ inches}} = \frac{10}{50 \text{ inches}} = \frac{0.2}{1 \text{ inch}}$$

$$t_{at10} = \frac{|N|}{R.v} = \frac{|10|}{(0.2 \text{ per inch})(100 \text{ inches per second})} = \frac{10}{\left(\frac{0.2}{1 \text{ inches}}\right)\left(\frac{100 \text{ inches}}{1 \text{ second}}\right)} = 0.5 \text{ second}$$

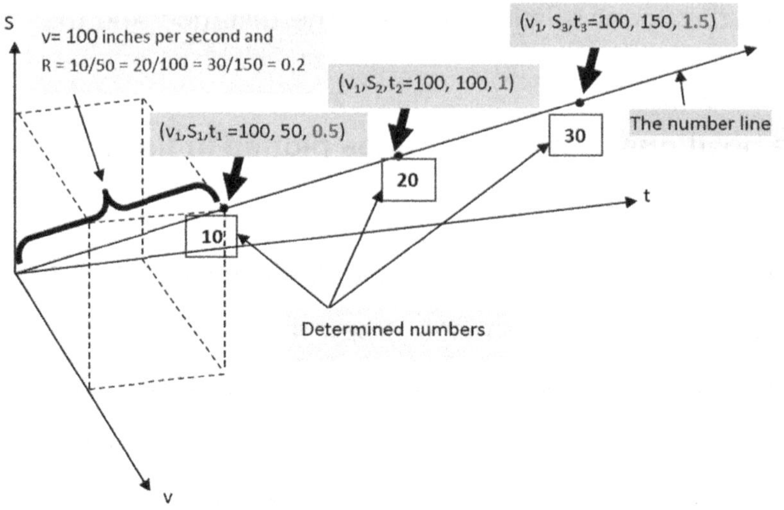

Figure 11: The five values, t, S, N, v, and R in each point are different from each other

Figure 11 shows "a position" at the number line where there are five values inside it with three- dimensional figure and **especially when time can be decimal.**

At N = 10, the distance S =50 (inches), the time value t = 0.5 (seconds), the line is written by using speed as $(S_2-S_1)/(t_2-t_1)$ = (50-0)/ (0.5-0) = 100 (inches per second). A change of the number to the distance and vice versa requires the use of value of R ($R = \frac{|N|}{S}$).

For exmaple, absolute of N = S.R or S = absolue of N / R.

In this case 10 = [50. (0.2)] or 50 = [absolute of (10)/0.2].

The time equation as one-inch equation explaining of many equations in physics and economics

Michio Kaku, theoretical American physicists, who mentioned about the existence of Theory of Everything (TOE) and one-inch equation that can explain all forces in physics. His prediction was written in his book “Beyond Einstein” in 1987, but nobody had found until now. The time equation shows that the existence of an equation explaining many equations in physics is real and well-founded. Moreover, the time equation can be used to explain all equations in economics as well. The equations of Newton, Einstein, and Planck are three examples in physics and consumption and saving equations in macroeconomics can be rewritten by using the time equation as follows: t = $\frac{F}{Rvg}$, t = $\frac{E}{Rvc^2}$, t = $\frac{E}{Rvh}$,

t = $\frac{(C-C_a)}{Rv(MPC)}$, and t = $\frac{(S+S_a)}{Rv(MPS)}$ respectively.

The principle of using time equation to explain all equations

All of equations in physics and economics consist of two main parts: dependent and independent variables. Usually, independent variable will be determined as horizontal axis that is a number line. The changing of the independent variable to be the determined number in time equation will be able to explain all equations by using time equation. The five equations can be explained by the time equation in the following subtopics.

Newton's equation and the time equation

Example 1, the time equation can be used to explain Newton's equation about gravity (F = mg) as follows:

Step I: Let us convert Newton's equation to a function:

F = f (m).

Step II: solve the independent variable as follows:

$F = 0 + (\frac{\Delta F}{\Delta m}).m$; $F = 0 + g.m$; as a result, $m = \frac{F}{g}$. See Figure 12.

Step III: The independent variable replaces the determined number in the time equation; as a result, the independent variable becomes the new form involved with the time equation as follows:

$\frac{|m|}{R} = S$; $S = vt$; then new form of m = Rvt and

$$t = \frac{m}{Rv} = m\left(\frac{1}{Rv}\right) = \frac{F}{Rvg}$$

Step IV: new form of Newton's equation: F = Rvtg.

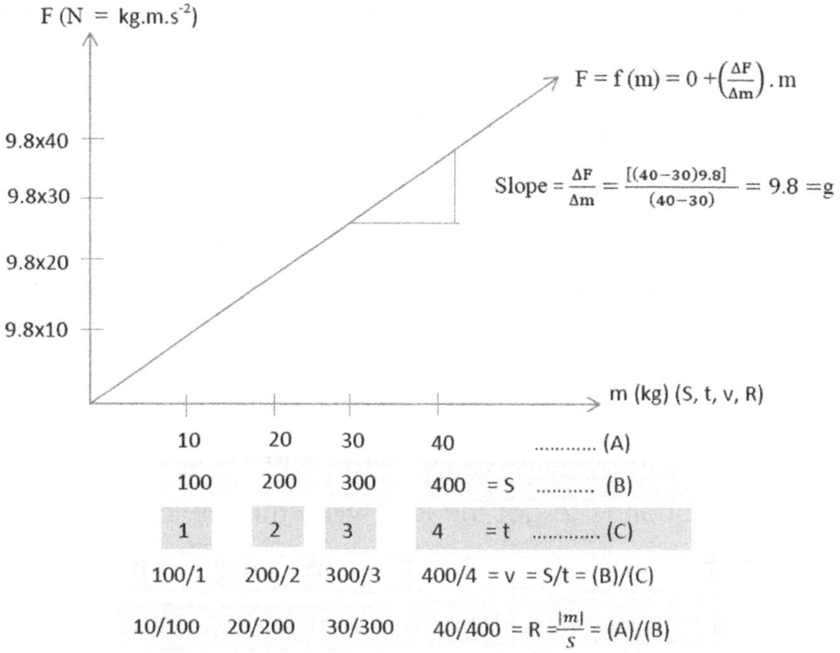

Figure 12: Time equation can explain Newton's equation.

In the Figure, the time equation is showed as the "equation of all equations". The five equations used include; (1) Newton's equation, (2) Einstein's equation, (3) Planck's equation, (4) the equation of Consumption in macroeconomics, and (5) the equation of saving in macroeconomics to demonstrate. Notably, mass (m), which is the independent variable in Newton's equation, is represented by N in the time equation. The m has other four values; S, t, v, and R. If we want to know the Newton's equation, we can use the time equation to explain.

In Figure 12, each number N is called “a determined number” because different drawers usually give different names for representing things they want to talk about. We always see the determined numbers in our own number lines. However, in a number line, there are at least four dimensions that we could be hidden on first assessment, but that can be calculated as follows: the distance S of the number from the origin, the ratio R between N (absolute value) and S, the velocity v that the drawer of the line uses to get the line, and the time value t that the drawer spends to get the line. With this method, only one value of time can be used to get mass (m) and force (F) and vice versa.

In this case

$$t = \frac{F}{Rvg}$$

$$3 = 9.8x30 \left(\frac{300}{30}\right)\left(\frac{3}{300}\right)\left(\frac{1}{9.8}\right)$$

$$t = \frac{m}{Rv}$$

$$3 = (30)\left(\frac{300}{30}\right)\left(\frac{3}{300}\right)$$

And $$m = Rvt$$

$$30 = \left(\frac{30}{300}\right)\left(\frac{300}{3}\right)3$$

$$F = Rvtg$$

$$9.8x30 = \left(\frac{30}{300}\right)\left(\frac{300}{3}\right)(3)\ (9.8)$$

Einstein's equation and the time equation

Example 2, the time equation, can be used to explain Einstein's equation ($E = mc^2$) as follows:

Step I: Let us represent Einstein's equation as a function:

$E = f(m)$.

Step II: solve the independent variable as follows:

$E = 0 + (\frac{\Delta E}{\Delta m}).m$; $E = 0 + c^2.m$; as a result, $m = \frac{E}{c^2}$.

See Figure 13.

Step III: the independent variable replaces the determined number in the time equation. As a result, the independent variable becomes the new form involved with the time equation as follows:

$\frac{|m|}{R} = S$; $S = vt$; then new form of $m = Rvt$ and

$$t = \frac{m}{Rv} = m\left(\frac{1}{Rv}\right) = \frac{E}{Rvc^2}$$

Step IV: New form of Einstein's equation by using the time equation: $E = Rvtc^2$.

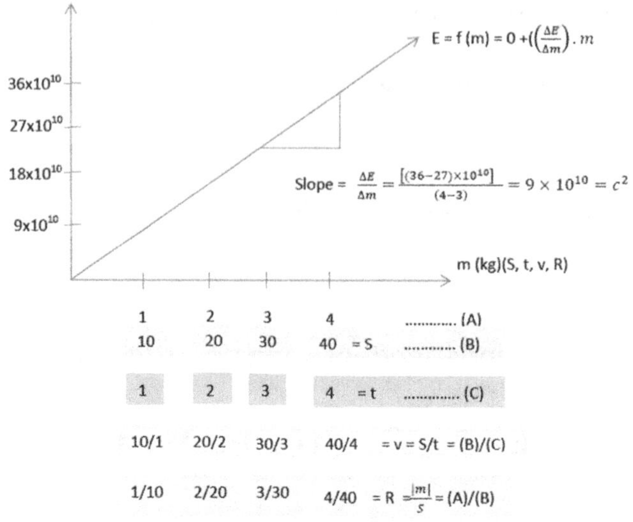

Figure 13: The time equation can explain Einstein' s equation.

In the figure, the time equation is shown as the "equation of all equations". The Einstein's equation was used to demonstrate. Notably, the mass m (an independent variable in Einstein's equation) is shown as N in the time equation. The mass m comprises other four values such as S, t, v, and R.

In this case

$$t = \frac{E}{Rvc^2}$$

$$4 = 36x10^{10}\left(\frac{10}{1}\right)\left(\frac{1}{10}\right)\left(\frac{1}{9x10^{10}}\right)$$

$$t = \frac{m}{Rv}$$

$$4 = (4)\left(\frac{10}{1}\right)\left(\frac{1}{10}\right)$$

And $m = Rvt$

$$4 = \left(\frac{1}{10}\right)\left(\frac{10}{1}\right)4$$

$$E = Rvtc^2$$

$$36x10^{10} = \left(\frac{1}{10}\right)\left(\frac{10}{1}\right)(4)\,(9 \times 10^{10})$$

Planck's equation and the time equation

Example 3, the time equation can be used to explain Planck's equation (E = h ν) as follows:

Step I: Let us represent Planck's equation as a function:

E = f (ν)

Step II: solve the independent variable as follows:

$E = 0 + (\frac{\Delta E}{\Delta \nu}).\ \nu$; $E = 0 + h\,\nu$; as a result, $\nu = \frac{E}{h}$. See Figure 14.

Step III: The independent variable replaces the determined number in the time equation. As a result, the independent variable becomes the new value in the time equation as follows:

$\frac{|\nu|}{R}$ = S; S = vt; then new form of ν = Rvt and

$$t = \frac{\nu}{Rv} = \nu\left(\frac{1}{Rv}\right) = \frac{E}{Rvh}$$

Step IV: New form of Planck's equation: E = Rvth.

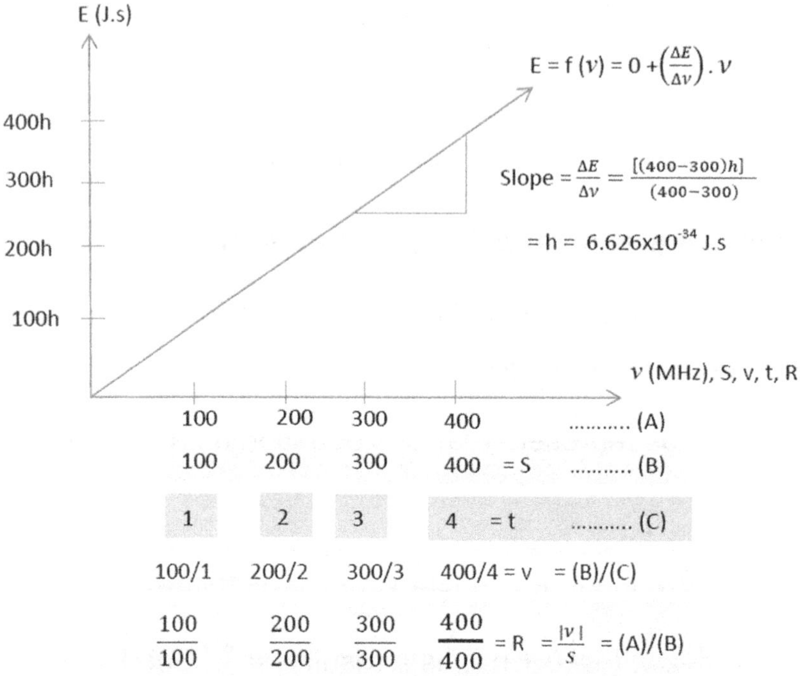

Figure 14: The time equation can explain Planck's equation, E = h ν.

Figure 14 shows how time equation is the equation of all equations. The Planck's equation was used to demonstrate. The time equation can be used to explain the value of Planck's equation as shown above. Notably, frequency ν, which is an independent variable in Planck's equation, is shown as N in the time equation. The value ν has other four values inside it: S, t, v, and R.

In this case

$$t = \frac{F}{Rvh}$$

$$2 = 200xh\left(\frac{100}{100}\right)\left(\frac{2}{200}\right)\left(\frac{1}{h}\right)$$

$$t = \frac{v}{Rv}$$

$$2 = (200)\left(\frac{200}{200}\right)\left(\frac{2}{200}\right)$$

And $$m = Rvt$$

$$200 = \left(\frac{200}{200}\right)\left(\frac{200}{2}\right)2$$

$$E = Rvth$$

$$200xh = \left(\frac{200}{200}\right)\left(\frac{200}{2}\right)(2)\,(h)$$

$$200x6.626x10^{-34} = \left(\frac{200}{200}\right)\left(\frac{200}{2}\right)(2)\,(6.626x10^{-34})$$

Consumption equation in macroeconomics and the time equation

Example 4, the time equation can be used to explain consumption's equation [$C = C_a + (MPC)Y_d$] in macroeconomics as follows:

Step I: Let us represent the consumption's equation as a function: $C = f(Y_d)$. See Figure 15.

Step II: solve the independent variable as follows:

$C = Ca + (\frac{\Delta C}{\Delta Yd})$. Y_d; $C = Ca + (MPC)\ Y_d$; as a result, $Y_d = \frac{C - C_a}{MPC}$.

Step III: The independent variable replaces the determined number in the time equation; as a result, the independent variable becomes the new form involved with the time equation as follows:

$$\frac{|Y_d|}{R} = S;\ S = vt;\ \text{then new form of } Y_d = Rvt \text{ and}$$

$$t = \frac{Y_d}{Rv} = Y_d\left(\frac{1}{Rv}\right) = \frac{C - C_a}{Rv(MPC)}$$

Step IV: The new form of the consumption's equation:

C = Ca + (MPC) Rvt.

In this case

$$t = \frac{C - C_a}{Rv(MPC)}$$

$$3 = (3{,}400\text{-}1{,}000)\left(\frac{30}{3{,}000}\right)\left(\frac{3}{30}\right)\left(\frac{1}{0.8}\right)$$

$$t = \frac{Y_d}{Rv}$$

$$3 = (3{,}000)\left(\frac{30}{3{,}000}\right)\left(\frac{3}{30}\right)$$

And Y_d = Rvt

$$3{,}000 = \left(\frac{3{,}000}{30}\right)\left(\frac{30}{3}\right)3$$

C = Ca + (MPC) Rvt

$$3{,}400 = 1{,}000 + (0.8)\left(\frac{3{,}000}{30}\right)\left(\frac{30}{3}\right)(3)$$

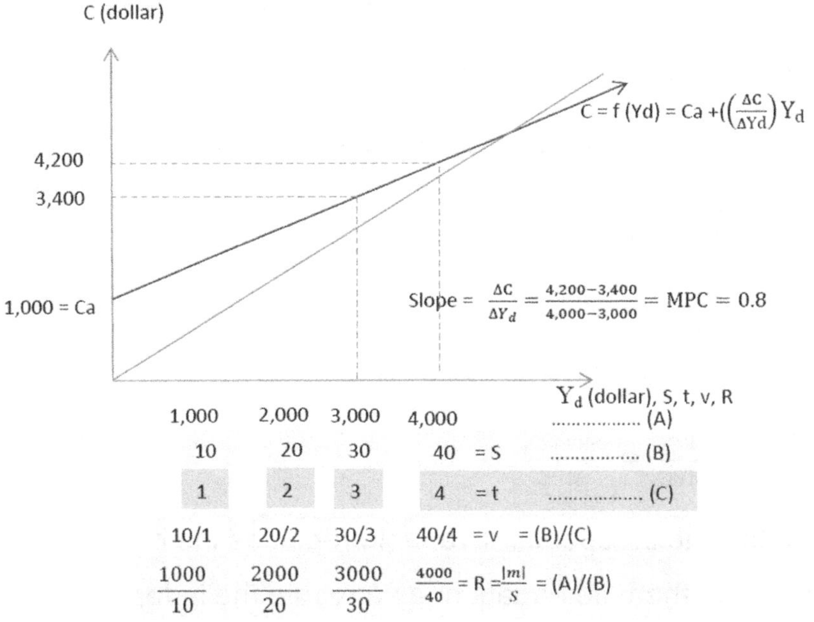

Figure 15: The time equation can explain the consumption equation in macroeconomics.

In the figure above, time equation is shown as the equation of all equations. In this case, disposable income Y_d (an independent variable in the equation of consumption) is represented by N. The Y_d has four values; S, t, v, and R. If we want to know the equation of consumption, we can use the time equation to explain.

Saving equation in macroeconomics and the time equation

Example 5, the time equation can be used to explain saving's equation [Sav = -Ca+ (1-MPC) Y_d = -Ca + (MPS).Y_d] in macroeconomics as follows:

Step I: Let us represent saving's equation as a function:

Sav = f (Y_d). See Figure 16.

Step II: solve the independent variable as follows:

$$Sav = -C_a + \left(\frac{\Delta C}{\Delta Y_d}\right).\ Y_d$$

$$= -C_a + (1\text{-}MPC)\ Y_d$$

$$= -C_a + (MPS)\ Y_d\text{; as a result,}$$

$$Y_d = \frac{Sav+C_a}{1-MPC} = \frac{Sav+C_a}{MPS}.$$

Step III: The independent variable replaces the determined number in the time equation; as a result, the independent variable becomes the new form related with the time equation as follows:

$\frac{|Y_d|}{R}$ = S; S = vt; then new form of Y_d = Rvt and

$$t = \frac{Y_d}{Rv} = Y_d\left(\frac{1}{Rv}\right) = \frac{Sav+C_a}{Rv(MPS)}$$

Step IV: The new form of saving's equation:

Sav = -Ca + (MPS) Rvt.

In this case

$$t = \frac{Sav + C_a}{Rv(MPS)}$$

$$5 = (0+1{,}000)\left(\frac{30}{3{,}000}\right)\left(\frac{3}{30}\right)\left(\frac{1}{0.2}\right)$$

$$t = \frac{Y_d}{Rv}$$

$$5 = (5000)\left(\frac{50}{5{,}000}\right)\left(\frac{5}{50}\right)$$

$$\text{And } Y_d = Rvt$$

$$5{,}000 = \left(\frac{5{,}000}{50}\right)\left(\frac{50}{5}\right)5$$

$$Sav = -Ca + (MPS)\,Rvt$$

$$0 = -1{,}000 + (0.2)\left(\frac{5{,}000}{50}\right)\left(\frac{50}{5}\right)(5)$$

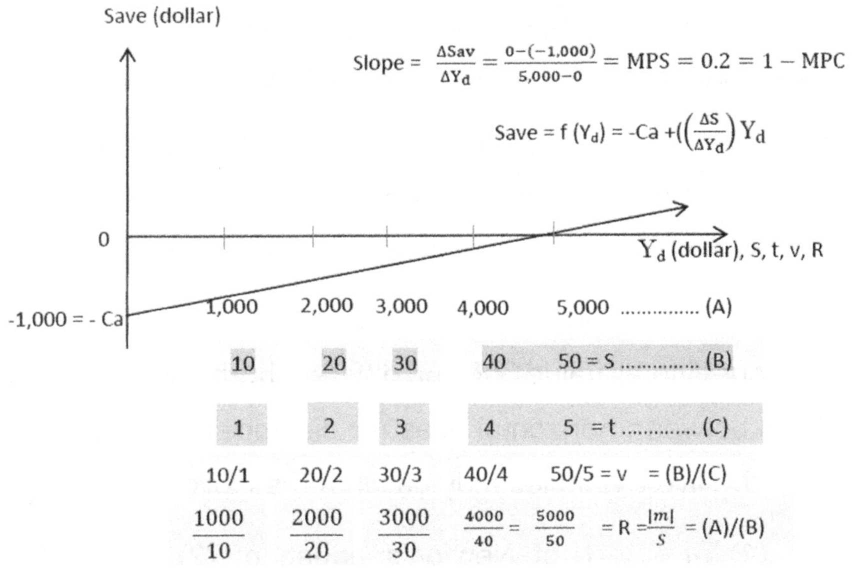

Figure 16: The time equation can explain the saving equation in macroeconomics.

In the figure, the disposable income Y_d, which is an independent variable in the equation of saving, is shown as N in the time equation. The Y_d has other four values inside it: S, t, v, and R. If we want to know the equation of saving, we can use the time equation to explain.

The five examples are the evidence of the time equation used for solving of many equations in physics and economics.

The time equation as the "king equation" used for solving N equations at one time

Normally, the N unknown variables can be solved if we have N equations. However, the time equation may help us solve the complicating problems with N variables in a second and may be new ways of solving mathematical problems that link to develop other disciplines especially physics and economics because the time equation has shown that we can solve N variables simultaneously by using a value of time at one time.

In Figure 12 to 16, we can use the only one value of time such as t_3 for getting all values of x on different horizontal axes. The relation between horizontal axes (or distances) and values of time of the five examples may be written as follow:

> (1) m = g (t) of Newton's equation; (2) m = f (t) of Einstein's equation; (3) ν = h (t) of Planck's equation; (4) Y_d = i (t) of consumption equation, and (5) Y_d = j (t) of saving equation.

If we know the only two dimensions as follow: (i) the velocities in these functions, (ii) the ratio R, we can solve all functions with a value of time. For example, if we determine the value of time as $t = t_3$; as a result, we can receive the set of numbers matching with the time values on all horizontal lines as follow: (1) 30 of Newton's equation; (2) 3 of Einstein's equation; (3) 300 of Planck's equation; (4) 3,000 of consumption equation, and (5) 3,000 of saving the equation. By-product of t_3 is we can get the all five dependent variables as well.

For example, in Newton's equation, (1) F = f (m) = 30x9.8 unit; (2) in Einstein's equation, E = f (m) = 27×10^{10} unit; (3) in Planck's equation, E = f (ν) = 300h unit; (4) consumption equation, C = f (Y_d) = 3,400 unit; and (5) in saving equation, Sav = f (Y_d) = -400 unit. This concept can be used by mathematicians, physicists, and economists to solve million equations by using the only one value of time (Boonsri, 2002).

It is beneficial for scientists in the future including economists who are studying open system of economy that has many variables. Similarly, physicists always do their experiments with close system; so many equations in physics can be solved at one time with the only one value of time (Boonsri, 2016).

The application of this method, for examples, a computer that has a tiny size like a penny coin may be invented and robotic technology will be developed rapidly. The new future vehicles of transportation may be invented because physicists understand all four fundamental forces consisting of Strong, weak, electromagnetic, and gravitational forces very well and can use each of them for producing new cheap energy.

Computers with new displays "Air Screens", the dream of sci-fi filmmakers, will be true if the equation works for all. Our favorite singers performing their music in front of us on three-dimensional air screens in our living room will be invented. Professional surgeons will advise an intern working at a far hospital because they see inside the body of a client on the air screen. An entrepreneur will be getting more money because of faster and accurate information for making decision from an equation instead of many equations.

We can also present N number lines occurring simultaneously. There are two cases of velocities of drawing of N number lines: (1) the same speeds of drawing of N lines that can show a spherical shape like fireworks at night and (2) the different speeds of drawing. We can explain mathematically these two cases as follows:

Given that t_1 is a value of time matching with the position of a determined number in a number line no.1. Similarly, t_N is a value of time matching with the position of a determined number in a number line no.N.

$$t_1 = \frac{s_1}{v_1}, \quad t_2 = \frac{s_2}{v_2}, \quad t_3 = \frac{s_3}{v_3}, \quad t_4 = \frac{s_4}{v_4}, \quad t_5 = \frac{s_5}{v_5},$$

$$t_6 = \frac{s_6}{v_6}, \quad t_7 = \frac{s_7}{v_7}, \quad t_8 = \frac{s_8}{v_8}, \quad t_9 = \frac{s_9}{v_9}, \quad t_{10} = \frac{s_{10}}{v_{10}},$$

$$t_{11} = \frac{s_{11}}{v_{11}}, \quad t_{12} = \frac{s_{12}}{v_{12}}, \quad t_{13} = \frac{s_{13}}{v_{13}}, \quad t_{14} = \frac{s_{14}}{v_{14}}, \quad t_{15} = \frac{s_{15}}{v_{15}},$$

$$t_{16} = \frac{s_{16}}{v_{16}}, \quad t_{17} = \frac{s_{17}}{v_{17}}, \quad t_{18} = \frac{s_{18}}{v_{18}}, \ldots\ldots, \quad t_N = \frac{s_N}{v_N}.$$

And in the first case: $v_1 = v_2 = v_3 = v_4 = v_5 = v_6 = v_7 = v_8 = v_9 = v_{10} = v_{11} = v_{12} = v_{13} = v_{14} = v_{15} = v_{16} = v_{17} = v_{18} = v_{19} = ... = v_N$. See Figure 17.

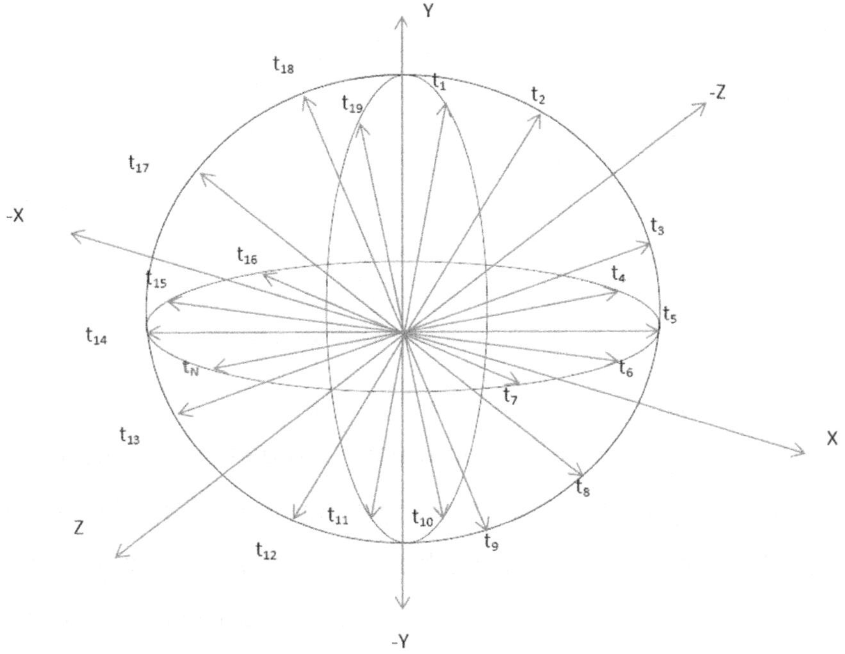

Figure 17: The solution of N equations with only one time value, t_1 to t_N.

In the above figure, all number lines have many numbers. Thus, if all number lines occurr simultaneously, at the same speeds of drawing, then they look like a firework exploding in the sky. Each number has other four values inside it: S, t, v, and R. Further, all number lines consisting of all independent variables occurring at the same speeds of drawing, are shown as the spherical shape like fireworks.

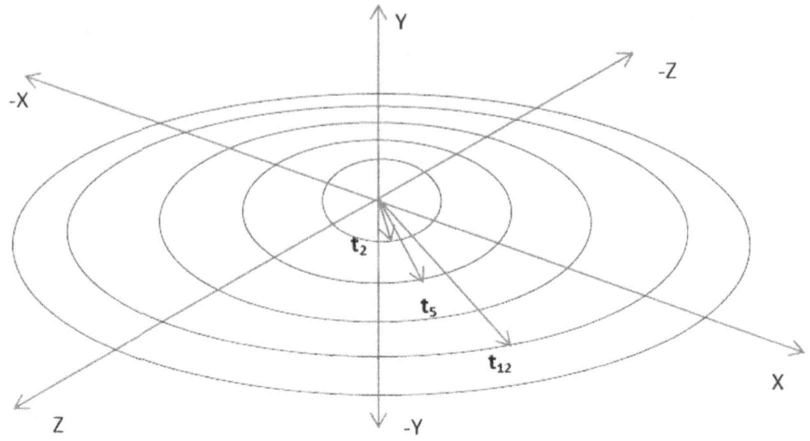

Figure 18: All number lines occurring simultaneously with three different speeds of drawing.

The figure also represents time equation as the equation of all equations. All the number lines have many numbers. However, in each number, there are other four values inside it, which include S, t, v, and R. In addition, all number lines consisting of all independent variables occurring at the different speeds of drawing are shown as the different lines with different lengths. If we want to know both the independent and dependent variables of all equations, then a value of time to solve all equations at one time can be used. Otherwise, the determination of the velocity and R of all equations are the known standards to determine the variables.

In the second case, if all number lines occur simultaneously with the different velocities of drawing, they can be explained mathematically as follows:

$v_1 \neq v_2 \neq v_3 \neq v_4 \neq v_5 \neq v_6 \neq v_7 \neq v_8 \neq v_9 \neq v_{10} \neq v_{11} \neq v_{12} \neq v_{13} \neq v_{14} \neq v_{15} \neq v_{16} \neq v17 \neq v18 \neq v19 \neq \ldots \neq v_N$. See Figure 18.

The basic concept of drawing multidimensional figures

Physicists explain a point as the zero-dimensional object with no any expansion in space ("Point (geometry)", 2016; "geometry facts, information, pictures | Encyclopedia.com articles about geometry", 2016). In the vector concept, the point is the only one or none of the vectors, and if these vectors have been set together, a plane area would be created. In addition, until now, scientists especially mathematicians believe that they can draw a graph with three dimensions in maximum.

We draw N number lines passing the origin, and values of time in the N number lines are matching with the positions of numbers, so the origin as zero may be zero-dimensional object or many dimensional objects that can show with many numbers or many values of time.

Normally, when we look at a picture, we thought that the picture has two dimensions; however, we always overlook the value of time that the picture was taken. Time is in everything. Time links all dimensions together. Each picture has two dimensions and one dimension of time not two dimensions. In other words, each picture has at least three dimensions at the moment that the picture was taken. In addition, each picture has its identity. It is unique. You can notice that you could not take the same details of all dimensions in your photos again.

Scientists think that time is the fourth dimension. In my opinion, I agree and disagree with them simultaneously. Time is the fourth dimension if we think that time does not change. In contrast, time is the fourth, fifth, sixth, seventh, and so on if we think that nothing is the same in every second. Imagine that we are standing at a space station looking at the earth travelling around the sun and we have been taking pictures of the earth in every second for five minutes. When we see the first, the second, and the third picture, we will realize that nothing is the same.

If we think that all dimensions are different from each other, we will realize that every moment we are living in new dimension of a value of time. Each value of time is different from each other. Everything has been changing all the time, so each value of time can be counted as different dimensions. Normally, we always see things continuously, so we do not realize that nothing is the same or we overlook the change of all dimensions. Everything is older and older. Everything is changing. For example, after every picture had been taken, it was older and older in every moment. Each value of time has its own unique dimensions. At $t = 1$, a man is married man, and at $t = 2$, the man is a father.

Similarly, the time dimension in all dimensions can open the new view of the point. For example, when we are standing on the highest building in a city and looking at the horizon of the earth, we will see the horizon as the straight line. However, the horizon is not a straight line, it is a curve. See Figure 19.

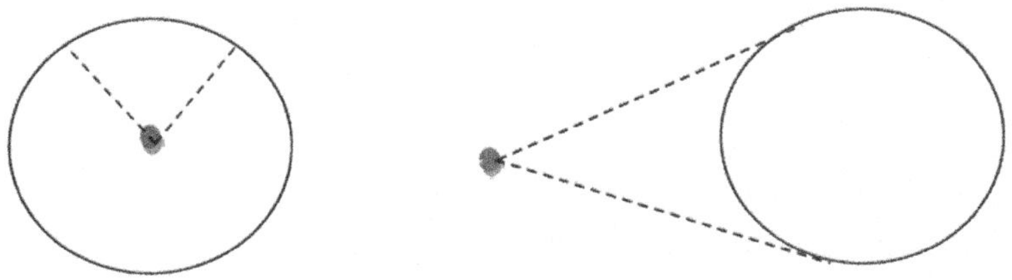

Figure 19: The red point represents the point we are standing and looking at something. If we are looking at the horizon of the earth, we will see the curve as a straight line.

In addition, if you are looking at the edge of a piece of paper; it looks like a straight line as well. See Figure 20. However, the width and the length of the paper are hidden inside the angle you are looking. From this point, if the time is increasing, then the space may be increasing. Notice that the line may be an area. See Figure 21. If we are looking a coin with cross section, it looks like a circle without thickness. However, the thickness of the coin is not seen at the place we are looking as well. From this point, if the time is increasing, then the space is increasing as well. Notice that the area may be a volume, which has values of time matching with numbers. See Figure 22. (These phenomena support the discovery of Edwin Hubble in 1929 about the expansion of our universe.)

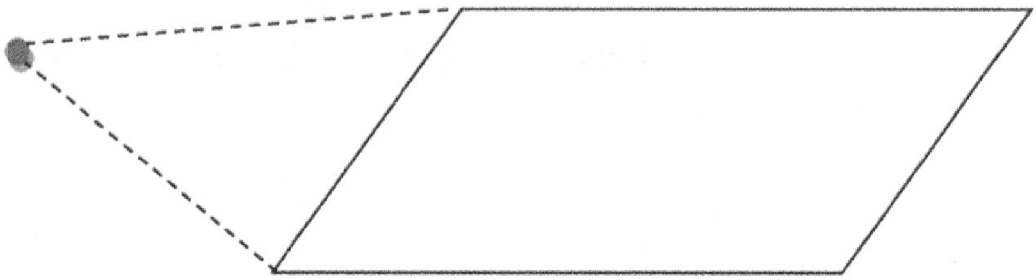

Figure 20: The area of a piece of paper is seen as a straight line as well. The red point represents the standing point that we are looking at.

The figure shows the hidden dimensions when we see things. Imagine we are looking at the only edge of a piece of paper, we will be able see it as a thin line (the length). Otherwise, it is not possible to see its width or its area. In this case, the hidden dimensions are its width and its area.

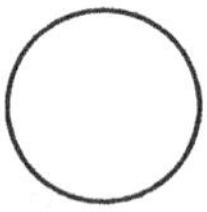

Figure 21: Coins looked at with a view of cross section

The figure shows the hidden dimensions when we see things. Looking at edge of a tower of coins would only allow us to the object as a circle (an area). Besides, it is not possible to see its thickness. In this case, the hidden dimensions represent its thickness and its volume.

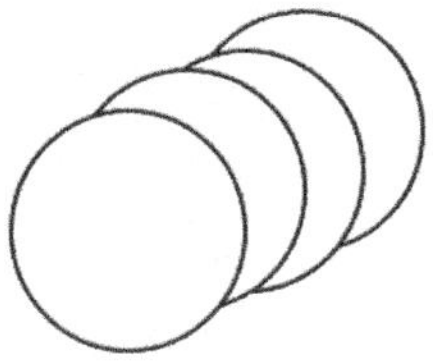

Figure 22: Coins looked at with a side view

The figure shows thickness and its volume of the object shown in *figure 21*.

From the new views of a point, points, time and space, we can develop our view to look things. For example, in the topic of prices of perfect competition in long run in microeconomics, the point A in two dimensions in the figure 23 may be drawn in many dimensions in the figure 24. The new view of a point may help us get the basic concept of drawing of multidimensional figure in the near future (Boonsri, 2015).

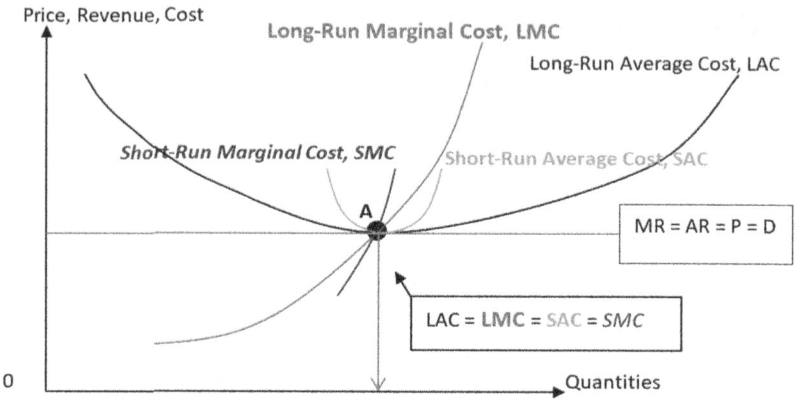

Figure 23: The point has eight dimensions. It is not a zero-dimensional object.

In the figure, there are eight dimensions in the point A that act as the evidence that a point is not a zero-dimensional object following the present mathematicians.

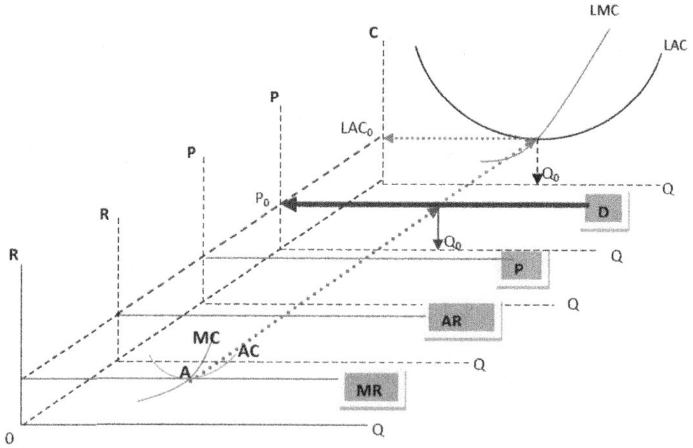

Figure 24: The point A in Figure 23 may be explained by the basic concept of multidimensional figure

Figure 24 is an expansion of figure 23, where all dimensions can be seen clearly. This presentation is the basic of how to draw the figure that has many dimensions more than three dimensions. (Right now, scientists believe we can draw only three-dimensional figure).

Minkowski's Light Cone model and Light spherical shape

In the previous chapter, we referred to Hermann Minkowski, who used his 4-D space-time model to describe time dimension as the fourth dimension separating from three principal axes. With his model, the path of time dimension with mathematical model, (x, y, z, t), indicated that the positions of the values of time locates in space among three principal axes. Right now, physicists are using his model to explain some events occurring in space concerning time dimension.

The time equation shows that values of time always match with the determined numbers x, y, and z. In other words, all numbers that mathematicians are using have their own values of time. We could not separate values of time from all numbers in a number line. In addition, we cannot use Cartesian form as (x, y, z, t) because when we use Cartesian form as (x, y, z), time values have already been in all numbers x, y, and z. In other words, Minkowski's model, which separates the time dimension from three principal axes, is imaginary.

With the Cartesian as (x, y, z, t), we can use partial derivative with respect to time because it is reasonable. However, the time equation indicated that when a value of time changes, the values of x or y or z will change automatically and simultaneously. Mathematicians are supposed to be consider this point to determine the changing of the time dimension while x, y, and z are constant.

The Minkowski's Light Cone model is imaginary as well. With Minkowski's model, time can run with negative values in past Light Cone, and with positive values in future Light Cone. With the time equation, the values of time from the origin are always positive as same as distance, and the value of time at the origin is zero. Minkowski thought that time as negative because he thought that negative determined numbers having negative time values as well.

With Minkowski's Light Cone model, the light in space has a shape as a cone. Minkowski determined the horizontal axis for explaining the space dimension and the vertical axis for explaining the time dimension. A photon moving between two axes for explaining the Light Cone leads to the light ray as well. Nevertheless, the light in space in nature has spherical shape because of these reasons: (1) we can draw N number lines passing the origin, the values of time in all lines, (2) N photons always move with the same speeds from the source of the light like a firework exploding. Imagine that if so and so shoots a firework into the sky, when the firework explodes, the light will appear at the exploding point and will expand around the point in all directions. People around the point, the left and the right, the upper and the bottom, the front and the back, can see the light simultaneously.

The new characteristics of time indicated that a value of time is in everything including a number in a number line. This characteristic supports Einstein's Relativity theory because time in the theory related to the velocity of the objects moving with the speed of light. The theory shows that time cannot be separated from three principal axes. The time value is the cause and distance is the result in equation (1) in this chapter. The difference between time and space can be showed by time equation as well. The time dimension cannot be decreased, but space can be decreased. While anyone is traveling with the speed of light or over the speed of light, the values of time that mankind spends for travelling are still positive.

Summary of the chapter

In the first chapter, mathematicians use numbers for representing things in the universe, and the time dimension is in everything in the universe. In the second chapter, the positions of values of time matching with numbers in number lines were proved mathematically after the hypothesis about time in the first chapter had been created.

Before the time equation was found, the Newton's relation among distance, velocity, and time values had been used. The first result as the equation (1) showed the distance as a function of the value of time. If the time value increases, then the distance increases as well. The distance is the only positive value; so the time values could not be negative value as well.

After the equation (1) had been proved, the time equation (3) was found in the next step. The time equation (3) shows the relation between a determined number and a time value. It can be used for us to indicate the exact position of a value of time matching with a position of a determined number on a line. Each number has its own value of time; and it is unique.

The numbers are called “determined numbers” because they can be determined by different drawers in different ways to represent tangible dimensions such as weight and quantity, and intangible dimensions such as force, energy, and electricity.

Moreover, the determined numbers can be positive and negative values, but the values of time matching with the numbers and distance are positive. The Minkowski’s 4-D model that is imaginary because we could not separate time from numbers.

Time is in everything including in numbers that mathematicians used to represent everything, the big and the small worlds. Physicists always explain many forces, energy and mass with mathematical equations. The existence of an equation that has the characteristics following Kaku’s commending in physics (Kaku, 2016) is real and well founded;

however, the time equation can explain not only equations in physics but also in economics.

The time equation can be solves N equations with a value of time at one time. This new paradigm shift in mathematics will lead scientists to invent new innovation concerning mathematics especially computer and innovation concerning computer technology.

The time equation indicated that a point can be zero or N dimensions. The new view of the point showed that the basic concept of drawing of N-dimensional figures is possible. The drawing of N-dimensional figures may help our children in the future to see the worlds of science and social science different from us in present.

These most crucial parts of this chapter can be concluded at the following.

Hypothesis 1: Are values of time in numbers?

Mathematical proof 1= accept or

S = f (t) = values of time are in a line

Proof 2 = accept or

the positions of values of time in any lines can be calculated by time equation

The positions of values of time = using these two formulas:

$$t = \frac{S}{V} = \left(\frac{|N|}{R.V}\right).$$

Hypothesis of Michio Kaku: Is there one-inch equation explaining equations in physics?

Proof 3 = accept Hypothesis of Michio Kaku or

time equation explaining science equations

(Time equation can explain economic equations too.)

New knowledge

The time equation can solve N equations at one time by using a value of time.

The basic concept of drawing multidimensional figure.

Chapter 3
The Time Vectors

"Wisdom is not a product of schooling, but of the lifelong attempt to acquire it."

Albert Einstein

The next characteristics of time after Einstein's work have been found since 1999 step by step. When having tried to study about time, I was in a situation called paradigm shift. With the law of paradigm shift, I would be pushed at the starting point of the new paradigm. No mathematicians except Minkowski in the past talked about the relation between numbers and the values of time; however, the knowledge of time of Minkowski's 4-D model was not beneficial enough for me to answer my question on my discovery.

The existence of time in numbers presented in this book is big different from Minkowski's 4-D model. I have found that values of time always match with numbers in three axes which may be written in mathematical form as (x_t, y_t, z_t) instead of (x, y, z, t) of Minkowski's model. In this chapter, the new characteristics of time will be discussed.

Absolute and relative values

The difference between the absolute value and the relative value is about the frame of reference. What is the frame of reference? If we draw a line, the origin of the line is zero that is also a frame of reference. The frame of reference is used as standard for the other points. For example, if the first number as 10 is placed at the distance as 10 inches long from the origin, then with the same standard for measuring, the second number as 20 will be placed at the distance as 20 inches long from zero. The zero is always an absolute value but 10 and 20 are relative numbers. With Einstein's Relativity Theory, all numbers are relative numbers except zero (Boonsri, 2016).

The second example, we are driving 40 miles per hour. The speed of the car is a relative number because we are using ground of the earth as zero and the frame of reference of this measuring. If we change the velocity of the ground from zero to 1,670 miles per hour because the earth is orbiting around the sun, the new speed of the car with the second reference will be changed from 40 to 1,710 miles per hour, 1,670 + 40 =1,710. All numbers that we are measuring have to have their own frames of reference. If we change the value of any frames of reference, we will get the new values of numbers comparing to the frames of reference as well. In case of racing, the speed of two cars has to measure by using a frame of reference.

With Einstein's Special Relativity theory, the values measured without their frames of reference may be useless because we misunderstood the real meaning of them. For example, if someone is saying that his kids get grade one in a subject in a class; most of listeners may think the kids are weak in the subject. However, in some universities in the Philippine, the excellent grade of a class is one; this means that his kids are excellent students in the class. In contrast, if the lowest grade in the class is zero, it means that his kids are weak in the subject in the class. These examples showed that the listeners have to know well what the frame of reference of the grade is. The important of the frame of reference is one of the fundamental concepts of Einstein's Special Relativity Theory.

Normally, when we are measuring the velocity of our cars such as 90 kilometers per hour, we may not realize that we are using the speed of the ground as zero; that is, the frame of reference of this measuring. From this point, this sentence is true in the only case that the ground on the earth has its speed as zero. However, this sentence is untrue if the ground of the earth changes its speed from zero to others. For example, the earth is orbiting around the sun. The sun is orbiting around the center of galaxy. The galaxy is orbiting around the center of the universe. In addition, the center of our universe is moving, and nothing in the universe has the real speed of zero. In the universe, everything is changing.

If we want to measure any values in the universe, we have to know at least two important aspects: (1) what the frame of reference of the measurement, and (2) what value that we are measuring is. While we are trying to measure something, we usually have to determine something as the frame of reference as zero for measuring as well.

The absolute value, in contrast, is measured without any frames of reference; as the result, the value that people who is measuring is the same in all places in the universe. When we draw a line, the absolute value in the line is normally an origin, and the relative values are all numbers except the origin.

The value of time matching with the origin is the frame of reference and an absolute value simultaneously. In addition, the values of time at the other positions from the origin are relative values. The values of time are one, two, and three are relative values. Each time value in the line has its own position as well.

The time directions along the three principal axes

We can draw N number lines passing the origin of the three-dimensional body. Similarly, there are also N time values matching with numbers in the N number lines. See Figure 25.

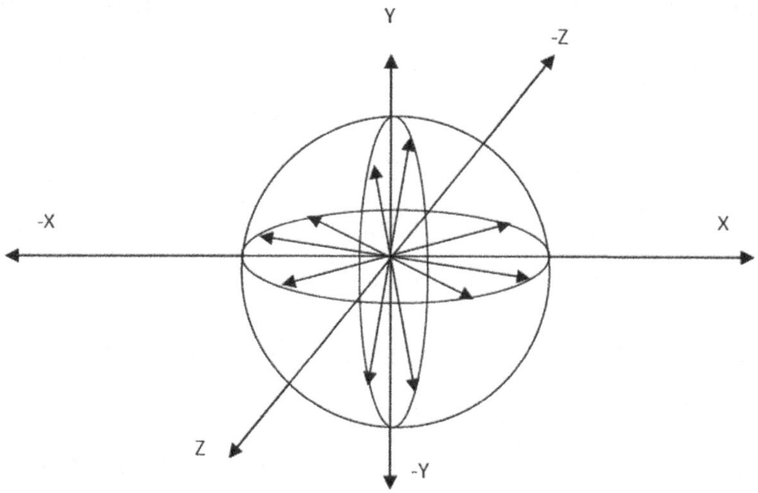

Figure 25: The N number lines can be drawn to pass an origin. The N time directions are along the lines as well.

The figure shows that there are N directions of time in three-dimensional body because we can draw N number lines in a three-dimensional body. Each number line has each direction of time.

Six directions of time along three axes

In general, a three-dimensional body has three principal axes. The values of time are matching with the positions of determined numbers in three axes as well. The value of time at the origin of the principal axes is absolute. In addition, the other values of time along the three principal axes always are relative. In the three principal axes, there are three pairs of two directions of the values of time. In the principal x-axis, there are the positive and the negative directions as x and –x. I named these directions "right-hand and left-hand sides." Similarly, in the principal y-axis, there are the positive and the negative directions as y, and –y. I named these directions "upper and bottom sides." In the principal z-axis, there are the positive and the negative directions as z, and –z. I named these directions "front and back."

In these six directions, we can find the exact positions of the values of time by using the time equation (3) as well. However, all values of time are only positive in both positive and the negative numbers.

Eight parts of space of a three-dimensional body

If we use three principal axes to divide the three-dimensional body into parts, then we will get eight parts of space. I name each part "an octodrant" with the symbol "oc." For example, if the symbol oc_1 is written for readers to read, this means the first part of the eight parts that I am referring to. Each part has its own ordered triple. The first part, oc_1, has its own ordered triple as (x, y, z). Similarly, the second to the eighth parts, oc_2 to oc_8, also have their ordered triples as follow:

(x, y, -z), (-x, y, -z), (-x, y, z), (x, -y, z), (x, -y, -z), (-x, -y, -z), and (-x, -y, z) respectively (Boonsri, 2002). See Figure 26.

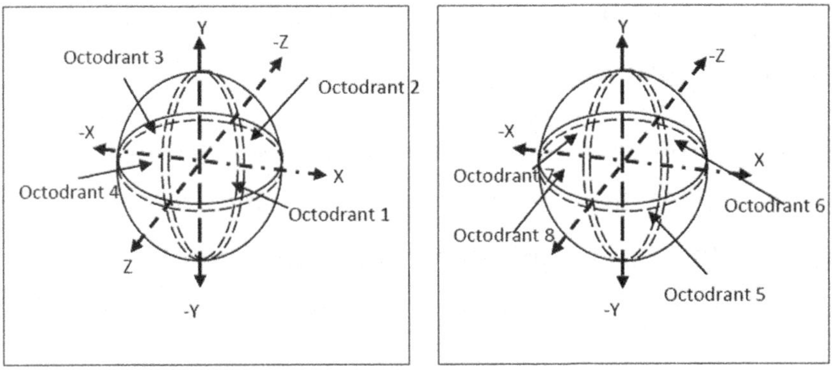

Figure 26: Eight parts of a three-dimensional body

In the figure above, the three–dimensional body can be divided into eight parts. Each of part I is called "an octodrant." There are eight parts (eight octodrants) in 3-D body. (Scientists use many ways to place the z-axis, but I use the z-axis to divide the space of the front and the back in this book.)

Time as a vector not a scalar

In three-dimensional body, we found that three principal axes divided the space into eight parts. If we can find the resultant time directions and the time values of all parts simultaneously, then we will conclude that the time dimension has the characteristic of a vector instead of a scalar.

First of all, at the first part of the three-dimensional body, the order triple in this part is (x, y, z), this means that the space in this part is the area among three principal axes with the positive number x, the positive number y, and the positive number z. See Figure 27 and 28. We have to use the equation

(3) to prove mathematically where the exact time values in three principal axes are.

At the determined number one on the x-axis without unit, if the distance from the origin is one inch long, the velocity of drawing this axis is one inch per second, and then the time value matching with the position of the determined number is one second. We can calculate the time value on the x-axis as follow:

The first method by using S and v;

$$t_x = \frac{S_x}{v_x} = \frac{1 \text{ inch}}{1 \text{ inch per second}} = \frac{1 \text{ inch}}{\frac{1 \text{ inch}}{1 \text{ second}}} = 1 \text{ second}$$

The second method by using N, R, and v;

$$R_x = \frac{|N_x|}{S_x \text{ inches}} = \frac{1}{1 \text{ inch}} = \frac{1}{1 \text{ inch}}$$

$$t_x = \frac{|N_x|}{R_x.v_x} = \frac{1}{(1 \text{ per inch})(1 \text{ inch per second})}$$

$$= \frac{1}{\left(\frac{1}{1 \text{ inch}}\right)\left(\frac{1 \text{ inch}}{1 \text{ second}}\right)} = 1 \text{ second}$$

At the determined number on the y-axis as one without unit, if the distance from the origin is one inch long, the velocity of drawing this axis is one inch per second, and then the time value at the same position of the determined number is one second. We can calculate the time value on the y-axis. See the proof in the part of Mathematical Proof of chapter 3.

Similarly, at the determined number on the z-axis as one without a unit, if the distance from the origin is one inch long, the velocity of drawing this axis is one inch per second, and then the time value at the same position of the determined number is one second. We can calculate the time value on the z-axis. See the proof in the part of Mathematical Proof of chapter 3.

To find the resultant direction of time of the part, we have to do three steps. At the first step, the first arrow has to be drawn from the origin to a determined number one along the x-axis; as a result, the direction of the positive x has just been drawn. **Next,** at the head of the first arrow, the second arrow with the same distance along the x-axis the running along a line parallel to the y-axis has to be drawn; as a result, the direction of the positive y has just been drawn. **Finally,** at the head of the second arrow, the third arrow with the same distance along the x-axis running along a line parallel to the z-axis has to be drawn; as a result, the direction of the positive z has just been drawn. In other words, the result direction of time is drawn from the origin to the head of the third arrow.

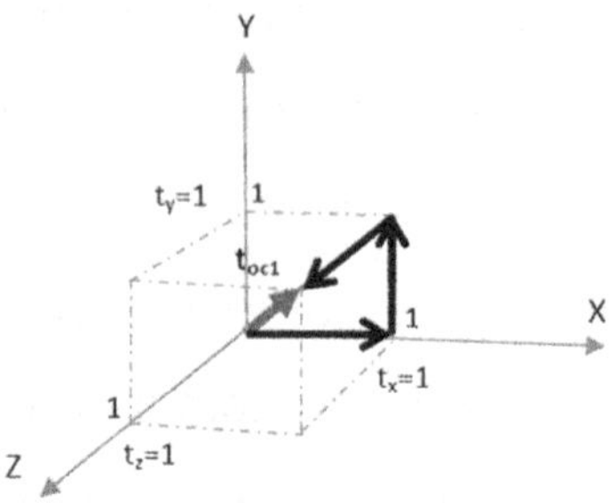

Figure 27: Ordered triple of the first part of a three-dimensional body is (x, y, z).

The figure shows the three steps (on the x-axis, y-axis, and z-axis) to get the resultant direction of time in first part. From the figure, the resultant direction of the values of time in the first part comes from running right, up, and front.

In Figure 28, the resultant direction of time in Figure 27 can be drawn again as a line from the origin to the point A_{oc1} at the corner of the square box as well.

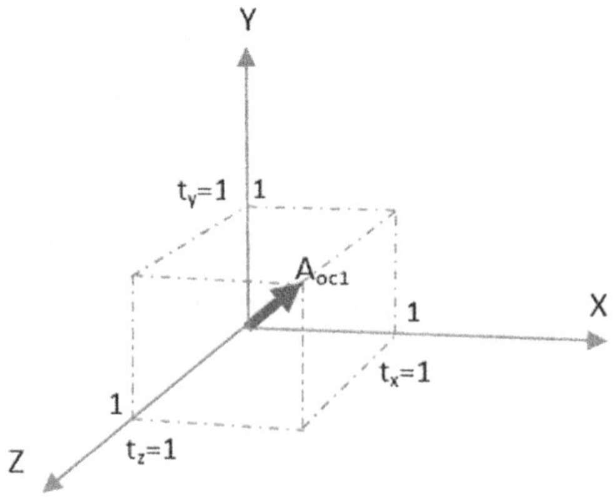

Figure 28: Figure 27 presented again with an arrow in a square box.

This is the representation of figure 27 again, but in this case, what is shown is only the resultant direction of time in first part (first octodrant).

The vector $\mathbf{A}_{oc1}$, the resultant direction of time of the first part, is made by three values of time in the three principal axes. I called these drawings in short “running right, up, and front.”

The vector $\mathbf{A}_{oc1}$ can be written in the form of mathematical vector as $\mathbf{A}_{oc1} = A_1\mathbf{i} + A_2\mathbf{j} + A_3\mathbf{k}$.

In addition, the magnitude of the vector $\mathbf{A}_{oc1}$ can be written mathematically as follow:

$$A_{oc1} = \sqrt{(A_1)^2 + (A_2)^2 + (A_3)^2}\ .$$

In this case, $A_1 = A_2 = A_3 = 1$, and then

$$A_{oc1} = \sqrt{(1+1+1)\,A_1^2} = \sqrt{3}\,A_1 = \sqrt{3}\,A_2 = \sqrt{3}\,A_3$$

$$= \sqrt{3}\,(1) = \sqrt{3}$$

If $A_1 \neq A_2 \neq A_3 \neq 1$, we will not use the square box to explain the resultant direction of time and the magnitude of the direction will not be $\sqrt{3}$.

Similarly, if the form of mathematical vector is

$\mathbf{A}_{oc7} = -A_1\mathbf{i} - A_2\mathbf{j} - A_3\mathbf{k}$, we can find the resultant direction of the vector $\mathbf{A}_{oc7}$ by drawing “running left, down, and back.”

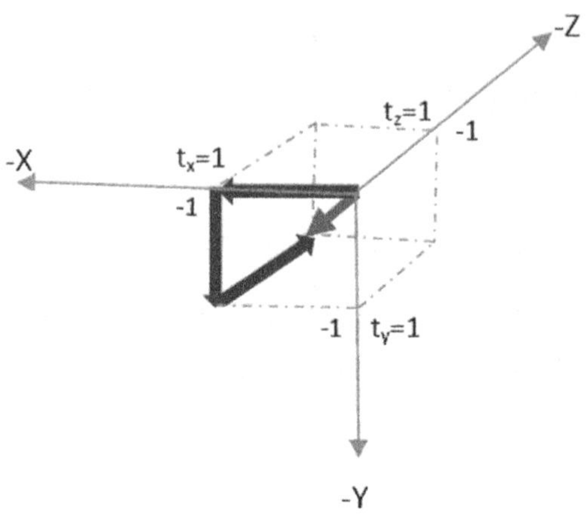

Figure 29: The resultant direction of the time values in the seventh part of the three-dimensional body, A_{oc7} or S_{oc7} or t_{oc7}, comes from running "left", "down" and "back."

In the figure, there are three steps (on the x-axis, y-axis, and z-axis) to get the resultant direction of time in seventh part (seventh octodrant).

In this case, the determined number A_1 is as long as A_2, A_3 and -1 or $A_1 = A_2 = A_3 = -1$.

The magnitude of the vector $\mathbf{A}_{oc7}$ in the seventh part can be written mathematically as follow:

$$A_{oc7} = \sqrt{(-A_1)^2 + (-A_2)^2 + (-A_3)^2}$$

$$= \sqrt{(A_1)^2 + (A_2)^2 + (A_3)^2}$$

$$= \sqrt{(1+1+1)\,A_1^2} = \sqrt{3}\,A_1 = \sqrt{3}\,A_2 = \sqrt{3}\,A_3$$

$$= \sqrt{3}\,(1) = \sqrt{3}\,.$$

If $A_1 \neq A_2 \neq A_3 \neq 1$, we will not use the square box to explain the resultant direction of time and the magnitude of the direction will not be $\sqrt{3}$.

Symbols explaining time as a vector

The number in the each part can be written by using many symbols. For example, the positive number x can be written as A_1 that is regular symbol in vector mathematics (or $x = A_1$) because x and A_1 are the symbol of the positive number on the x-axis. The positive number x can be written as N_x that is a determined number represented in this book (or $x = N_x$) as well. The equation (2) can be written in the forms of x, A_1, and N_x as follow: $\frac{|x|}{R_x} = \frac{|A_1|}{R_x} = \frac{|N_x|}{R_x} = S_x = v_x t_x$.

From the equation (2), the positive number x can be rewritten as S_xR_x that is the symbol showing distance (or $x = S_xR_x$) because $|x| = x = S_xR_x$. The positive number x can be written as $v_xt_xR_x$ that is the symbol showing the value of time (or $x = v_xt_xR_x$) because $|x| = x = v_xt_xR_x$. See Figure 30.

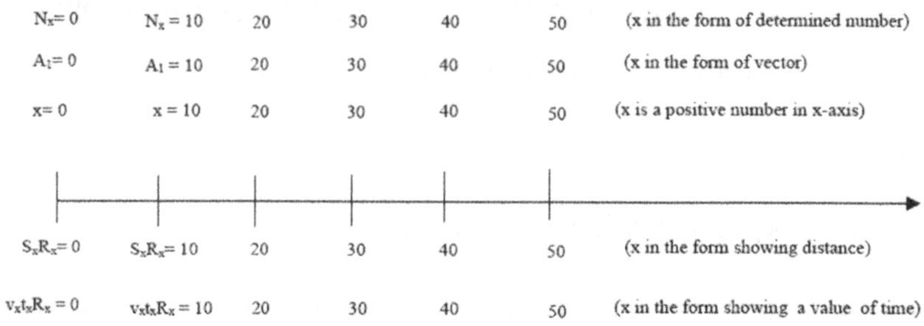

Figure 30: A number determined by a drawer as different symbols.

The figure shows that "a number" can be named as many symbols. Although, different symbols exist, all of them show a number. It looks like a man having many names that are called by parents, friends, his lovers, his enemies, his competitors. Each name has a specific aim to be called. For example, the symbol $S_xR_x = 10$ was created for showing the **positive** distance of the **positive** number from the origin. The symbol $v_xt_xR_x = 10$ was created to show there is a **positive** time value in the number.

In the negative case, the negative number -x can be written as $-A_1$ that is regular symbol in vector on the x-axis (or $-x = -A_1$). The negative number -x can be written as $-N_x$ that is the determined number represented in this book (or $-x = -N_x$). The negative number can be rewritten as $-(S_xR_x)$ that is the form showing distance.

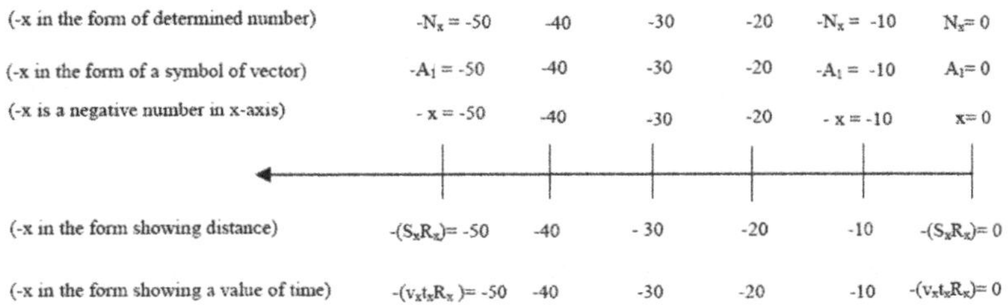

Figure 31: A number in x-axis written in different symbols.

The figure shows that "a number with negative value" can be named as many symbols. Although different symbols exist, all symbols show a specific number. It looks like a man having many names that he is referred to by his parents, friends, his lovers, his enemies, his competitors. Each name has a specific aim to be called. For example, the symbol – (S_xR_x) = -10 was created for showing the **positive** distance of the **negative** number from the origin. The symbol - $(v_xt_xR_x)$ = -10 was created to show there is a **positive** time value in the **negative** number.

Similarly, y, -y, z and –z can be rewritten by using previous reasons.

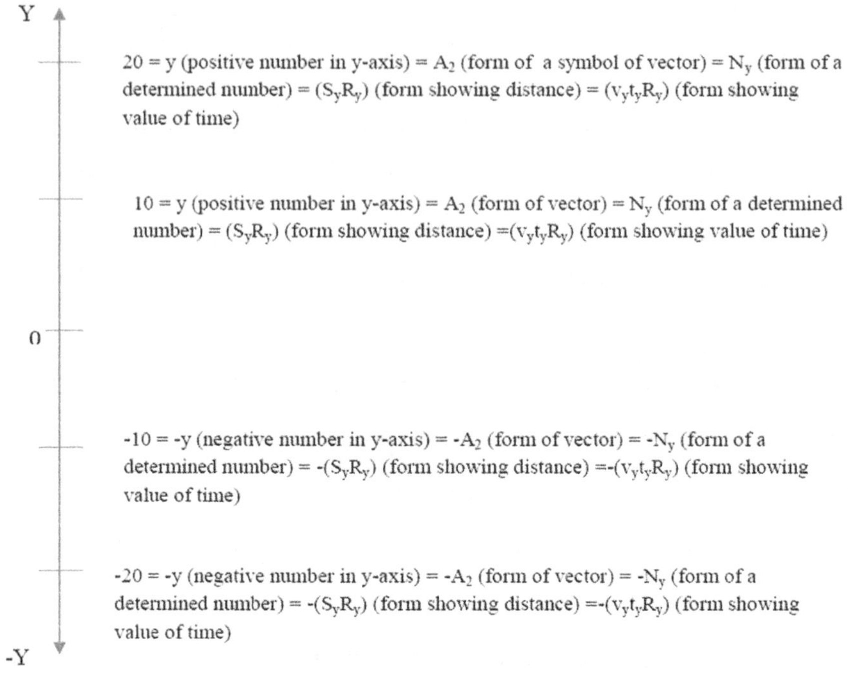

Figure 32: A number in y-axis written in different symbols

The symbol – (S_yR_y) = -10 was created to show the **positive** distance of the **negative** number from the origin in y-axis. The symbol - ($v_yt_yR_y$) = -10 was created to show there is a **positive** time value in the **negative** number in y-axis.

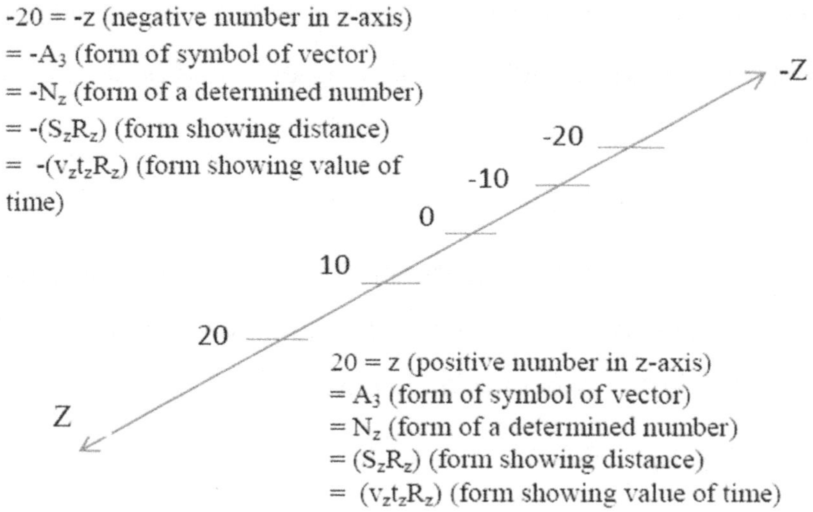

Figure 33: A number can be written in different symbols

The symbol – $(S_zR_z) = -10$ were created to show the **positive** distance of the **negative** number from the origin in z-axis. The symbol - $(v_zt_zR_z) = -10$ was created to show there is a **positive** time value in the **negative** number in z-axis.

And these are symbols used in this book as follow:

(1) $x = A_1 = N_x = S_xR_x = v_xt_xR_x$

(2) $-x = -A_1 = -N_x = -(S_xR_x) = -(v_xt_xR_x)$

(3) $y = A_2 = N_y = S_yR_y = v_yt_yR_y$

(4) $-y = -A_2 = -N_y = -(S_yR_y) = -(v_yt_yR_y)$

(5) $z = A_3 = N_z = S_zR_z = v_zt_zR_z$

(6) $-z = -A_3 = -N_z = -(S_zR_z) = -(v_zt_zR_z)$

If we change the vector $\mathbf{A}_{oc1}$ consisting of $A_1\,\mathbf{i} + A_2\,\mathbf{j} + A_3\,\mathbf{j}$ to the symbol showing the determined number in this book, we may rewrite the vector $\mathbf{A}_{oc1}$ in the new form as follow:

$$\mathbf{N}_{oc1} = N_x\,\mathbf{i} + N_y\,\mathbf{j} + N_z\,\mathbf{k}$$

If we change the vector $\mathbf{A}_{oc1}$ consisting of $A_1\,\mathbf{i} + A_2\,\mathbf{j} + A_3\,\mathbf{j}$ to the symbol showing the distance in this book, we may rewrite the vector $\mathbf{A}_{oc1}$ in the new form as follow:

$$\mathbf{S}_{oc1} = (S_xR_x)\,\mathbf{i} + (S_yR_y)\,\mathbf{j} + (S_zR_z)\,\mathbf{k}$$

If we change the vector $\mathbf{A}_{oc1}$ consisting of $A_1\mathbf{i} + A_2\,\mathbf{j} + A_3\,\mathbf{j}$ to the symbol showing the value of time in this book, we may rewrite the vector $\mathbf{A}_{oc1}$ in the new form as follow:

$$\mathbf{t}_{oc1} = (v_xt_xR_x)\,\mathbf{i} + (v_yt_yR_y)\,\mathbf{j} + (v_zt_zR_z)\,\mathbf{k}$$

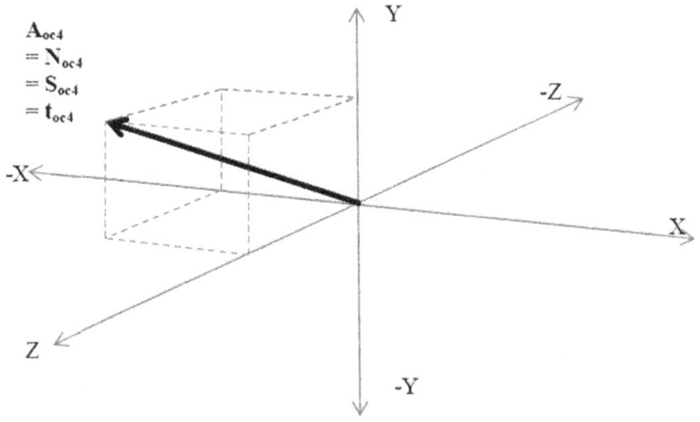

Figure 34: Resultant direction of time in the fourth part with different names

In the figure above, the resultant direction of time of the fourth part with different symbols (or names) is shown. The symbols;

- A-represents a mathematician who explains about vector in popular symbol
- N-means the determined number in the time equation
- S-is the distance in the time equation
- t is the symbol for the value of time as in the time equation.

The magnitude of the vector $\mathbf{A}_{oc1}$ can be written in the many forms: (1) the determined numbers, (2) the distances, and (3) the time values respectively as follow:

$$A_{oc1} = N_{oc1} = S_{oc1} = t_{oc1}$$

The resultant directions of time in the eight parts of three-dimensional body can be calculated and written into four various forms as follow: (1) the form of numbers such as $\mathbf{A}_{oc1}$ and $\mathbf{A}_{oc6}$, (2) the form of the determined numbers in this book such as $\mathbf{N}_{oc1}$ and $\mathbf{N}_{oc6}$, (3) the form of the distances such as $\mathbf{S}_{oc1}$ and $\mathbf{S}_{oc6}$, (4) the form of the values of time such as $\mathbf{t}_{oc1}$ and $\mathbf{t}_{oc6}$. The proof and all forms are shown in the part 'Mathematical Proof' of chapter 3.

The resultant directions of time of the eight parts are evidence of the time value as a vector not a scalar. These all resultant directions of the time values can be shown in Figures 35 to 42 as follow:

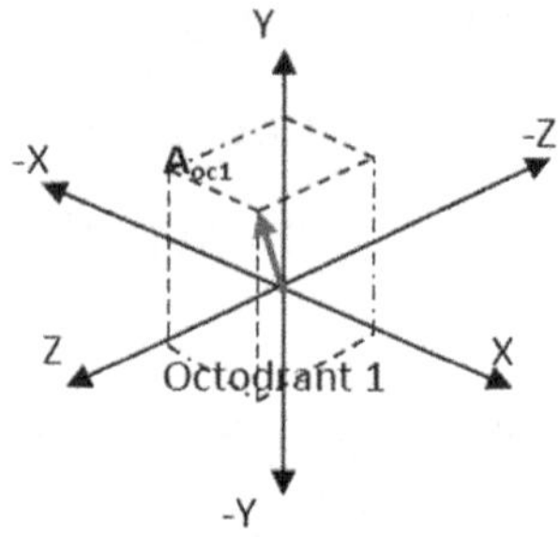

Figure 35: The ordered triple in the first part of the three-dimensional body is (x, y, z). The resultant direction of values of time, A_{oc1} or N_{oc1} or S_{oc1} or t_{oc1}, comes from running right, up, and front.

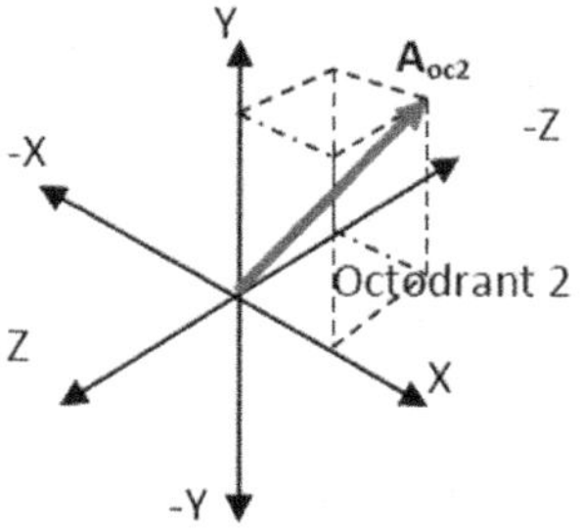

Figure 36: The ordered triple of the second part is (x, y, -z). The resultant direction of values of time, A_{oc2} or N_{oc2} or S_{oc2} or t_{oc2}, comes from running right, up, and back.

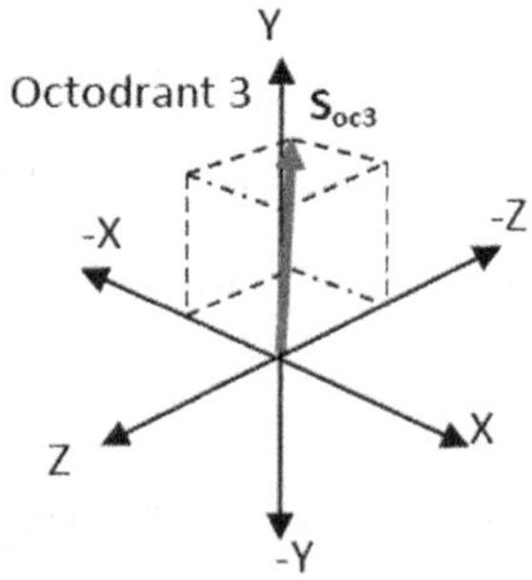

Figure 37: The ordered triple of the second part is (-x, y, -z). The resultant direction of time, A_{oc3} or N_{oc3} or S_{oc3} or t_{oc3}, comes from running left, up, and back.

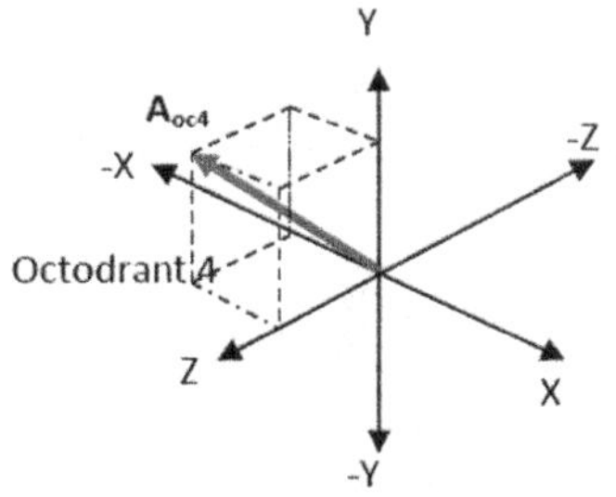

Figure 38: The ordered triple of the fourth part is (-x, y, z). The resultant direction of values of time, A_{oc4} or N_{oc4} or S_{oc4} or t_{oc4}, comes from running left, up, and front.

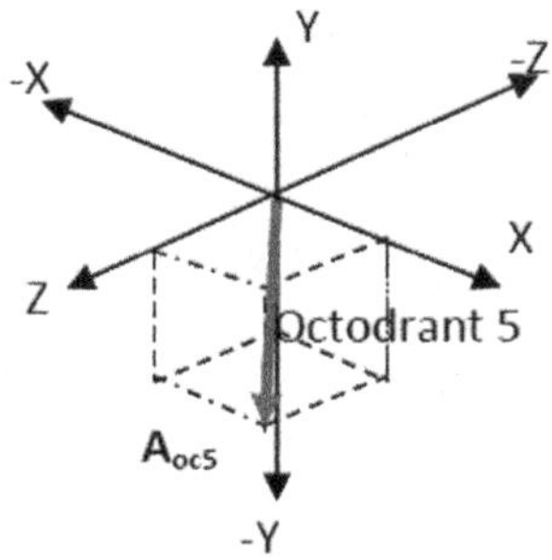

Figure 39: The ordered triple of the fifth part of the three-dimensional body is (x, -y, z). The resultant direction of values of time, A_{oc5} or N_{oc5} or S_{oc5} or t_{oc5}, comes from running right, down, and front.

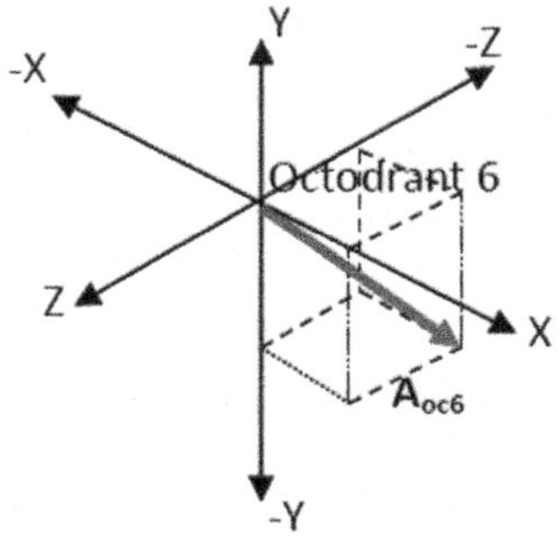

Figure 40: The ordered triple of the sixth part is (x, -y, -z). The resultant direction of values of time, A_{oc6} or N_{oc6} or S_{oc6} or t_{oc6}, comes from running right, up and back.

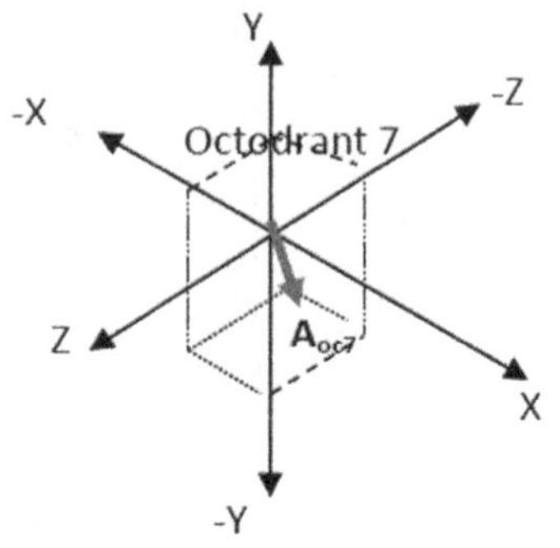

Figure 41: The ordered triple of the seventh part is (-x, -y, -z). The resultant direction of values of time, A_{oc7} or N_{oc7} or S_{oc7} or t_{oc7}, comes from running left, down, and back.

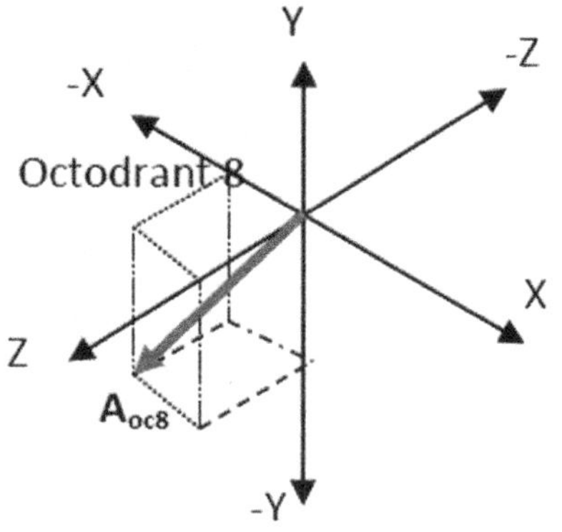

Figure 42: The ordered triple of the eighth part is (-x, -y, z). The resultant direction of values of time, A_{oc8} or N_{oc8} or S_{oc8} or t_{oc8}, comes from running left, down, and front.

Figures 35-42 show the resultant directions of time of the first to the eighth parts with different symbols (or names)

We found the resultant vectors of the values of time and the magnitudes of the vectors in the eight parts of the three-dimensional body simultaneously. This scientific evidence indicates that the time dimension has the new characteristics as a vector not a scalar (Boonsri, 2002).

Law of the direction of time

The time dimension has the characteristic of a vector not a scalar. This means that the direction of time is important which we have to consider about it. Having considered the direction of time, I found the law of the direction of time as follow: "if the different directions of time exist, the different values of time exist." In addition, time is in all dimensions; as a result, if the different values of time exist, then mass, energy, force, and field are different from the others. In other words, different directions of time can tell us many things such as different directions of force, different values of time, different characteristics such as different kinds of fields, different kinds of poles. In other words, each direction of time has its own characteristic or it is unique.

For example, the right direction of time on a number line shows the positive numbers that is different from the negative numbers. (If the different directions of time exist, the different values of time exist.)

If the end of the right time direction is the United States of America and the end of the left time direction is Thailand, then when USA is midnight, at the same time, Thailand is midday. (If the different directions of time exist, the different values of time exist.)

If the end of the right time direction is the right side of the California state and the end of the left time direction is the left side of that, then the time at the right side will be different from that at the left side. (If the different directions of time exist, the different values of time exist.)

If a traveler travels round the world, and he (or she) decides to travel to the east, he (or she) will get one day more in his (or her) life, but if he (or she) decides to travel to the west, he (or she) will lose one day more in his (or her) life. (If the different directions of time exist, the different values of time exist.)

If the end of the right time direction is the North Pole of a magnetic bar and the end of the left time direction is the South Pole, and then the magnetic force at the North Pole is different from the magnetic force at the South Pole of the bar. (If the different directions of time exist, the different characteristics of pole exist.)

If the end of the right time direction is the North Pole of the earth and the end of the left time direction is the South Pole, then when the North Pole is winter, at the same time, the South Pole is summer. (If the different directions of time exist, the different characteristics of pole exist.)

The different time directions in the field of logistics showed the different results of many values. A track can show many different time lines of logistics with different directions of logistic activities simultaneously. For example, while the track is running, many dimensions are increasing such as the value of goods, the wage of the driver driving the track, and the utility of customers, and the pollution, petroleum products used for the track running. (If the different directions of time exist, the different values of time exist.)

In contrast, at the same time, many dimensions are decreasing such as the quality of four wheels, the rest time for using the track, the ability of driver getting older, the freshness of the body of the track, the time of road for using. All activities occur simultaneously in the different time directions and make different results. Time is the linkage that links all activities of human beings (Kassakorn, 2014).

In general, we have different time directions. For example, we have the different birthday, different places, different families, different towns, different cities, different countries, different climate, and geography or continents. We are different from each other. We are unique because of the different time directions.

If the different time directions occur, then the different results will take place as well. In the other words, everything in economy and society has its own time direction. Everything in the universe is unique because of different time directions.

For example, taking a pill at different times, taking normal doze and over doze of medicine, taking medicine made from chemical substance and herb, these make the different results of healing. Mangoes plucked in and out the mango's season make different prices of selling. Prices of drink in a theater, prices of organic and inorganic vegetables, costs of goods with different transportations, costs of clothes with different materials, prices of meals at different times, these make different profits of sellers. The different groups of organization such as AFFA, NAFTA, and EU without Great Britain made the different results of solving economy problems.

The different time directions occur, the time values at different places in the earth occur as well. The different time directions occur, the time values at different sides of a window occur as well. The different time directions occur, the time values at different hands of a kid occur. The different time directions occur, the time values at different sides of a cell's wall happen. The different time directions occur, the time values at different sides of a particle occur too.

Summary of the chapter

The time equation (3) can be used to indicate the exact positions of the time values matching with the positions of numbers in a number line. We can draw N number lines passing the origin of three-dimensional body; this means that the time values are in all number lines as well. Three of N number lines of the three-dimensional body are three principal axes.

Three principal axes can be used to divide the space of the three-dimensional body into eight parts. The first part has its ordered triple as (x, y, z); and the second to the eighth parts have their ordered triples as (x, y, -z), (-x, y, -z), (-x, y, z), (x, -y, z), (x, -y, -z), (-x, -y, -z), and (-x, -y, z) respectively. Each ordered triple has its own time values matching with the positions of the determined numbers.

A number in three principal axes can be presented in many ways as follow: (1) using the regular numbers as (x, y, z) or symbols of vector (A_1, A_2, A_3), (2) using the determined number in this books (N_x, N_y, N_z), (3) using symbols showing distances (S_xR_x, S_yR_y, S_zR_z), (4) using the values of time with symbols ($v_xt_xR_x$, $v_yt_yR_y$, $v_zt_zR_z$).

With these symbols of numbers, (x, y, z) or (A_1, A_2, A_3) or (N_x, N_y, N_z), we could not see any time values, but the values of time always match with the positions of these numbers. Similarly, with these symbols concerning distances, (S_xR_x, S_yR_y, S_zR_z), we could not see any values of time in them too, but the values of time can be calculated by using the time equation (3). Finally, with these symbols showing the values of time, ($v_xt_xR_x$, $v_yt_yR_y$, $v_zt_zR_z$), the values of time can be shown.

In each part of the three-dimensional body, the resultant directions of time of all parts can be shown. No difference meaning among these symbols $\mathbf{A}_{oc1}$ = $\mathbf{N}_{oc1}$ = $\mathbf{S}_{oc1}$ = $\mathbf{t}_{oc8}$ took place. These are used to present the resultant directions of time of the first part in three-dimensional body with different symbols. These symbols were written with the bold characters. $\mathbf{A}_{oc1}$ to $\mathbf{A}_{oc8}$ or $\mathbf{N}_{oc1}$ to $\mathbf{N}_{oc8}$ or $\mathbf{S}_{oc1}$ to $\mathbf{S}_{oc8}$ or $\mathbf{t}_{oc1}$ to $\mathbf{t}_{oc8}$ are the resultant directions of time of the first to the eighth parts in three-dimensional body. The form of mathematical vector $\mathbf{A}_{oc1} = A_1\mathbf{i} + A_2\mathbf{j}+A_3\mathbf{k}$ expresses how to get the vector by running along the three directions of three principal axes. If we draw a line from the origin to the end point of the running, we will get the resultant direction of the time of all parts in three-dimensional body.

All vectors have their own magnitudes. Similarly, these four kinds of symbols are used to present the magnitudes of the resultant directions of time in eight parts of three-dimensional body: (1) A_{oc1} to A_{oc8}, (2) N_{oc1} to N_{oc8}, (3) S_{oc1} to S_{oc8}, and (4) t_{oc1} to t_{oc8}. For example, these different symbols, $A_{oc1} = N_{oc1} = S_{oc1} = t_{oc1}$, are the magnitude of resultant direction of time of the first part. The magnitude of the resultant direction of time of each part with different forms can be written as follow: $\sqrt{3}\,A_1 = \sqrt{3}\,A_2 = \sqrt{3}\,A_3 = \sqrt{3}\,N_x = \sqrt{3}\,N_y = \sqrt{3}\,N_z =$ $\sqrt{3}(S_xR_x) = \sqrt{3}\,(S_yR_y) = \sqrt{3}\,(S_zR_z) = \sqrt{3}\,(v_xt_xR_x) =$ $\sqrt{3}\,(v_yt_yR_y) = \sqrt{3}\,(v_zt_zR_z)$.

In eight parts of the three-dimensional body, we found eight resultant directions of time and the magnitudes of these directions. These are scientific evidence supporting time dimension as a vector not scalar.

The different directions of time indicate that everything in the universe is unique because it has its own direction of time.

Law of direction of time said that "If the different directions of time exist, the different values of time exist." This law of direction of time can be used to explain many different results of natural phenomena, activities of human beings, activities of logistics or economy.

Chapter 4
The Time Pyramid Inspired by Albert Einstein

You can't use an old map to explore a new world.

Albert Einstein

The characteristic of time in numbers and as a vector are the second and third characteristic of time respectively, after Einstein found the first characteristic of time in 1905. The next characteristic of time discussed in this chapter is time in magic angle, which is one of the most important topics. It is also the scientific evidence of existence of time in subatomic level. The evidence is that the new view at the subatomic level may help scientists to identify the secret in subatomic level by using time. The other secrete about time is the pyramidal shape, which links to control of chemical reaction in wanted and unwanted cells as described in the fifth chapter.

The six time-fields

The eight resultant directions of time in eight parts of a three-dimensional body were discussed in the third chapter. When having grouped these eight time directions, we will get six pyramidal shapes with a vertex that I named them "six time-fields (Boonsri, 2016)." See Figure 43.

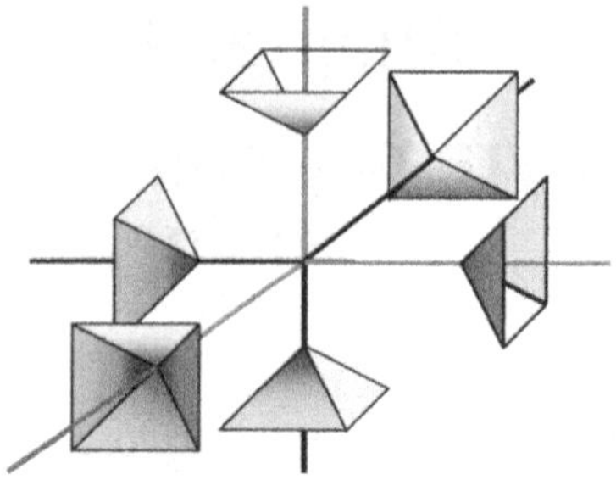

Figure 43: The name of a pyramidal shape in a three-dimensional body is "a time-field." There are at least six time-fields in the three-dimensional body.

The figure shows the six time-fields in 3-D body. Each time-field has the characteristic of a vector and not a scalar. However, this is just the theoretical conclusion, but in the fifth chapter, the conclusion by scientific experiments is discussed. Each time-field has its own resultant direction, but the same magnitude. In other words, these six time-fields have their specific resultant directions. In the next topic, how six time-fields were grouped will be discussed before the resultant directions of six time-fields are presented.

The first group consists of the resultant directions of time of the first, the second, the sixth, and the fifth parts of the three-dimensional body. We can write mathematically the first group with four different forms as follow: (1) the form of numbers as $\mathbf{A}_{oc1}$, $\mathbf{A}_{oc2}$, $\mathbf{A}_{oc5}$, and $\mathbf{A}_{oc6}$, (2) the form of the determined numbers in this book as $\mathbf{N}_{oc1}$, $\mathbf{N}_{oc2}$, $\mathbf{N}_{oc5}$, and $\mathbf{N}_{oc6}$, (3) the form of the distances as $\mathbf{S}_{oc1}$, $\mathbf{S}_{oc2}$, $\mathbf{S}_{oc5}$, and $\mathbf{S}_{oc6}$, (4) the form of the values of time as $\mathbf{t}_{oc1}$, $\mathbf{t}_{oc2}$, $\mathbf{t}_{oc5}$, and $\mathbf{t}_{oc6}$. I named the first group "the right time-field." See Figure 44.

The second group consists of the resultant directions of time of the third, the fourth, the seventh, and the eighth parts of a three-dimensional body with the form of the numbers as $\mathbf{A}_{oc3}$, $\mathbf{A}_{oc4}$, $\mathbf{A}_{oc7}$, and $\mathbf{A}_{oc8}$. I named the second group "the left time-field." See Figure 44.

The third group consists of the resultant directions of time of the first, the second, the third, and the fourth parts of a three-dimensional body with the form of the numbers as $\mathbf{A}_{oc1}$, $\mathbf{A}_{oc2}$, $\mathbf{A}_{oc3}$, and $\mathbf{A}_{oc4}$. I named the third group "the upper time-field." See Figure 45.

The forth group consists of the resultant directions of time of the fifth, the sixth, the seventh, and the eighth parts of a three-dimensional body with the form of the numbers as $\mathbf{A}_{oc5}$, $\mathbf{A}_{oc6}$, $\mathbf{A}_{oc7}$, and $\mathbf{A}_{oc8}$. I named the forth group "the bottom time-field." See Figure 45.

The fifth group consists of the resultant directions of time of the first, fourth, eighth, and the fifth parts of a three-dimensional body with the form of the numbers as $\mathbf{A}_{oc1}$, $\mathbf{A}_{oc4}$, $\mathbf{A}_{oc8}$, and $\mathbf{A}_{oc5}$. I named the fifth group "the front time-field." See Figure 46.

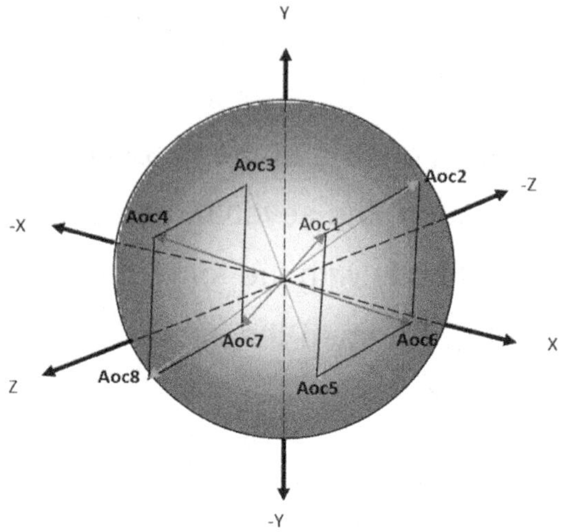

Figure 44: The right and the left time-fields.

The figure shows the right and the left time-fields in 3-D body.

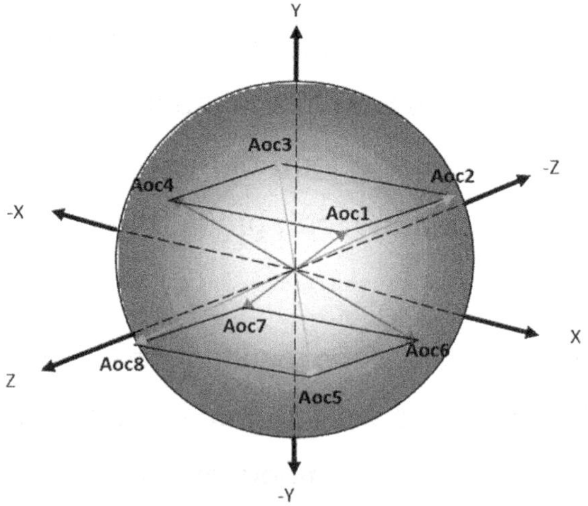

Figure 45: The upper and the bottom time-fields.

The sixth group consists of the resultant directions of time of the second, the third, the seventh, and the sixth parts of a three-dimensional body with the form of the numbers as $\mathbf{A}_{oc2}$,

$\mathbf{A}_{oc3}$, $\mathbf{A}_{oc6}$, and $\mathbf{A}_{oc7}$. I named the sixth group "the back time-field." See Figure 46.

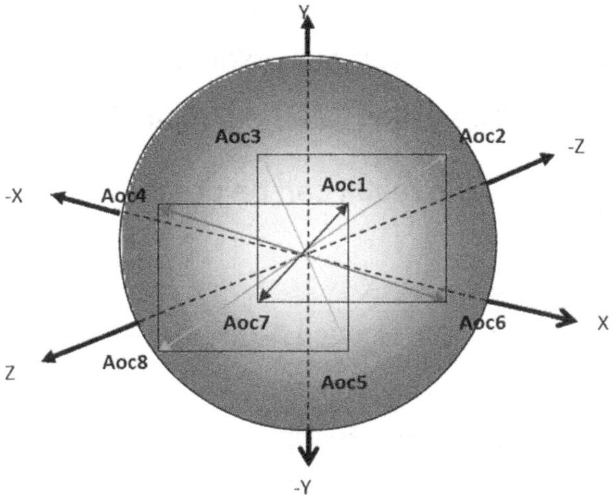

Figure 46: The front and the back time-fields.

The resultant directions of six time-fields

We can find resultant directions of six time-fields in three-dimensional body. Each time-field has the same size, but there is distinct resultant direction. The resultant direction of the time-field can be calculated by using four resultant directions of time in the four parts of the three-dimensional body. For example, if we want to get the resultant direction of the right time-field, we have to use four resultant directions of time in the four parts as follow: $\mathbf{A}_{oc1}$, $\mathbf{A}_{oc2}$, $\mathbf{A}_{oc5}$, and $\mathbf{A}_{oc6}$. In other words, four resultant directions of time in the first, second, fifth, and sixth parts of the three-dimensional body are the data used to calculate the resultant direction of the right time-field with the bold symbol of the vector as $\mathbf{A}_{PAright}$ or $\mathbf{A}_X$. The vector $\mathbf{A}_{PAright}$ has similar direction as the principal x-axis with the positive side, see Figure 49. In addition, the magnitude of the vector $\mathbf{A}_{PAright}$, with the symbol $A_{PAright}$ without boldness, can be calculated. See the proof in the section 'Mathematical Proof' of chapter 4.

The principle to get the result can be presented in the following chart.

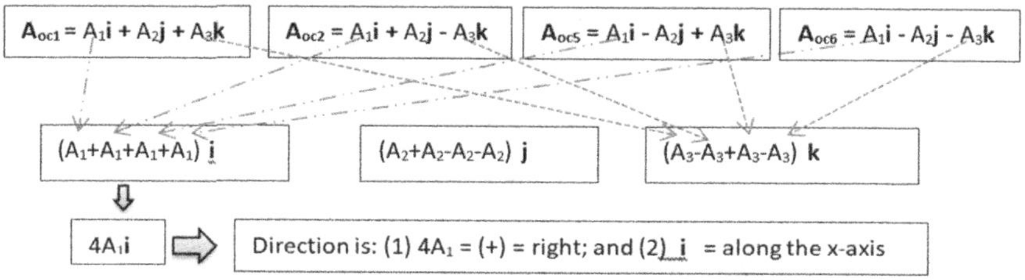

Figure 47: Chart of calculation of vectors

In the figure above, the equations show how to calculate the resultant direction of the right time-field in 3-D body and the interpretation of the results.

The principle to get the magnitude of the resultant direction can be presented in the following chart.

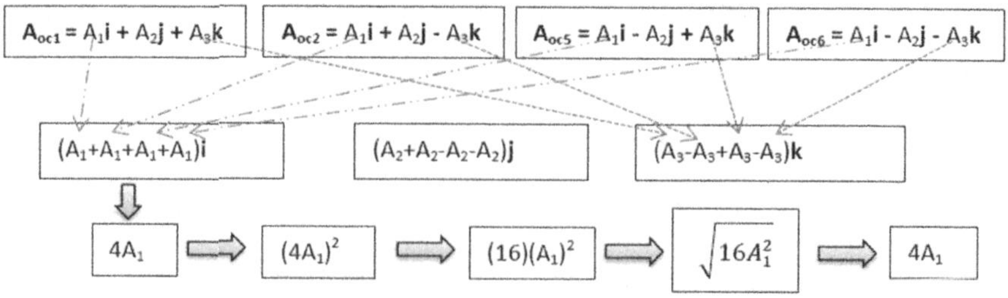

Figure 48: Chart showing calculation of magnitude of a vector

The calculations in the figure are for getting the magnitude of the resultant direction of the right time-field in 3-D body. Similarly, if we want to get the resultant direction of the left time-field, we have to use four resultant directions of time in the four parts as follow: $\mathbf{A}_{oc3}$, $\mathbf{A}_{oc4}$, $\mathbf{A}_{oc7}$, and $\mathbf{A}_{oc8}$. In other words, four resultant directions of time in the third, the forth, the seventh, and the eighth parts of the three-dimensional body are the data to calculate for getting the resultant direction of the left time-field with the bold symbol of the vector as $\mathbf{A}_{PAleft}$ or $\mathbf{A}_{(-X)}$. The vector $\mathbf{A}_{PAleft}$ is as same direction as the principal x-axis with the negative side. See Figure 49.

In addition, the magnitude of the vector $\mathbf{A}_{PAleft}$, the symbol as A_{PAleft} without boldness, can be calculated. The proof of these calculations is the part, 'Mathematical Proof' of chapter 4.

If we want to get the resultant direction of the upper time-field, we have to use four resultant time directions in the four parts: $\mathbf{A}_{oc1}$, $\mathbf{A}_{oc2}$, $\mathbf{A}_{oc3}$, and $\mathbf{A}_{oc4}$. In other words, four resultant directions of time in the first, the second, the third, and the forth parts of the three-dimensional body are the data to calculate for getting the resultant direction of the upper time-field with the bold symbol of the vector as $\mathbf{A}_{Paupper}$ or $\mathbf{A}_{Y}$. The vector $\mathbf{A}_{PAupper}$ is as same direction as the principal y-axis with the positive side. See Figure 49. In addition, the magnitude of the vector $\mathbf{A}_{PAupper}$, the symbol as $A_{PAupper}$ without boldness, can be calculated. The proof of these calculations is the part, 'Mathematical Proof' of chapter 4.

Similarly, if we want to get the resultant direction of the bottom time-field, we have to use four resultant directions of time in the four parts as follow: $\mathbf{A}_{oc5}$, $\mathbf{A}_{oc6}$, $\mathbf{A}_{oc7}$, and $\mathbf{A}_{oc8}$. In other words, four resultant time directions in the fifth, the sixth, the seventh, and the eighth parts of a three-dimensional body are the data to calculate for getting the resultant direction of the bottom time-field with the bold symbol of the vector as $\mathbf{A}_{Pabottom}$ or $\mathbf{A}_{(-Y)}$. The vector $\mathbf{A}_{PAbottom}$ is as same direction as the principal y-axis with the negative side. See Figure 49. In addition, the magnitude of the vector $\mathbf{A}_{PAbottom}$, the symbol as $A_{PAbottom}$ without boldness, can be calculated. The proof of these is in the part of Mathematical Proof of chapter 4.

If we want to get the resultant direction of the front time-field, we have to use four resultant time directions in the four parts as follow: $\mathbf{A}_{oc1}$, $\mathbf{A}_{oc4}$, $\mathbf{A}_{oc5}$, and $\mathbf{A}_{oc8}$. In other words, four resultant directions of time in the first, the forth, the fifth, and the eighth parts of the three-dimensional body are the data to calculate for getting the resultant direction of the front time-field with the bold symbol of the vector as $\mathbf{A}_{Pafront}$ or $\mathbf{A}_{Z}$. The vector $\mathbf{A}_{PAfront}$ is as same direction as the principal z-axis with the positive side. See Figure 49. In addition, the magnitude of the vector $\mathbf{A}_{PAfront}$, the symbol as $A_{PAfront}$ without boldness, can be calculated. The proof of these calculations is the part, 'Mathematical Proof' of chapter 4.

Similarly, if we want to get the resultant direction of the back time-field, we have to use four resultant directions time in the four parts as follow: $\mathbf{A}_{oc2}$, $\mathbf{A}_{oc3}$, $\mathbf{A}_{oc6}$, and $\mathbf{A}_{oc7}$. In other words, four resultant directions of time in the second, the third, the sixth, and the seventh parts of the three-dimensional body are the data to calculate for getting the resultant direction of the back time-field with the bold symbol of the vector as $\mathbf{A}_{Paback}$ or $\mathbf{A}_{(-Z)}$. The vector $\mathbf{A}_{PAback}$ is as same direction as the principal z-axis with the negative side. See Figure 49. In addition, the magnitude of the vector $\mathbf{A}_{PAback}$, the symbol as A_{PAback} without boldness, can be calculated. The proof is in the part 'Mathematical Proof' of chapter 4.

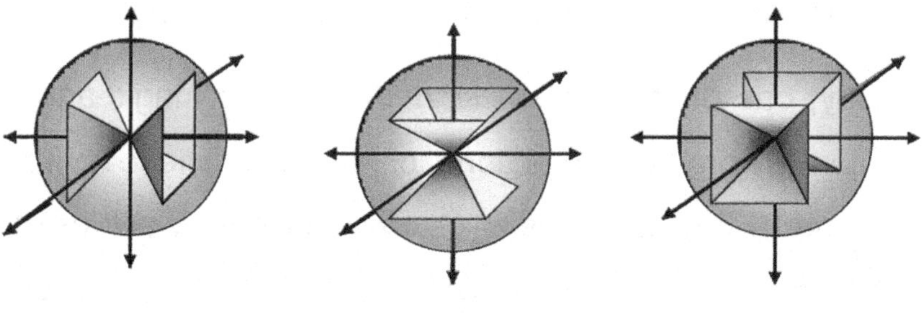

Figure 49: Three pairs of time-fields in a three-dimensional body.

The figure provides the mathematical proof and support of the six time-fields as vectors since their six resultant directions can be found.

The important angles of the time dimension in a three-dimensional body

In six time-fields in the three-dimensional body, we can find four important angles; ω, Φ, δ, and magic angle (γ). Present scientists know well that in the subatomic level, the magic angle exists. If we have the evidence of the existence of a value of time in the angle; then we will able to conclude that the time dimension is in subatomic level. In other words, the value of time in the magic angle is the evidence supporting the existence of time dimension in the three-dimensional body especially in subatomic level.

The angle of as 109.4712°

In general, an angle is the space between two lines that meet each other. When two lines meet at a point, an angle is formed. The two lines are called the arms of the angle. ("Angle", 2016.)

Similarly, one of the angles, ω, Φ, δ, and the magic angle γ, also come from two directions or vectors. These symbols consisting of cos ω or cos φ or cos δ or cos γ, are used to present these four angles called cosines Omega, Phi, Delta and Gamma respectively. Specifically, if I wrote $\cos\omega_{16}\ or\ \cos\varphi_{12}\ or\ \cos\delta_{13}$, it implies that the cosine angle Omega16, that is, the space between two resultant directions of time in the first and the sixth parts of the three-dimensional body. In addition, the cosine angle Phi12 is the space between two resultant time directions in the first and the second parts of the three-dimensional body, and the cosine angle Delta13 is the space between two resultant time directions in the first and the third parts of the three-dimensional body (Boonsri, 2016).

The resultant directions of time in the eight parts of the three-dimensional body in the form of numbers are the data to calculate for getting these angles. See the proof in the part of Mathematical Proof of chapter 4.

For example, we define the cosine angle Omega 16 between two vectors $\mathbf{A}_{oc1}$ and $\mathbf{A}_{oc6}$ by the formula as follow (Boonsri, 2003):

$$\cos \omega_{16} = \frac{\mathbf{A}_{oc1} \cdot \mathbf{A}_{oc6}}{\mathrm{A}_{oc1}.\,\mathrm{A}_{oc6}}$$

(Because $\mathbf{A}_{oc1} \cdot \mathbf{A}_{oc6} = \mathrm{A}_{oc1}.\mathrm{A}_{oc6} \cos \omega_{16}$)

The process of solving of the angle as example can be written in the following chart.

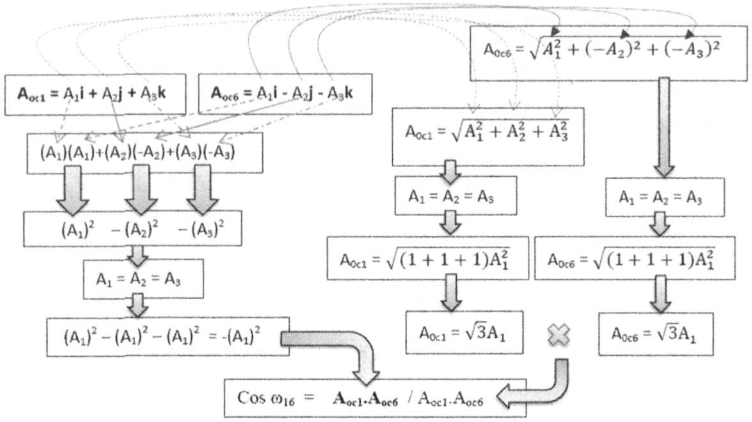

Figure 50: Chart showing process of calculation for getting an angle

The figure shows how to calculate to get the angle Cos ω_{16} in a time-field in 3-D body. The angle 109.4712°, was calculated by using two vectors as follow:

$\omega = \omega_{13} = \omega_{16} = \omega_{18} = \omega_{24} = \omega_{25} = \omega_{27} = \omega_{36} = \omega_{38} = \omega_{45} = \omega_{47} = \omega_{57} = \omega_{68} =$ 109.4712°. See Figure 51. See the proof in the part of Mathematical Proof of chapter 4.

We can find the magic angle as follow:

$$\omega = 109.4712° \therefore \frac{\omega}{2} = 54.7356° = \text{magic agle.}$$

I named this method "indirect method for getting magic angle." With indirect method for getting magic angles, we have twelve ways to solve this problem such as $\frac{\omega_{13}}{2}$ and $\frac{\omega_{68}}{2}$.

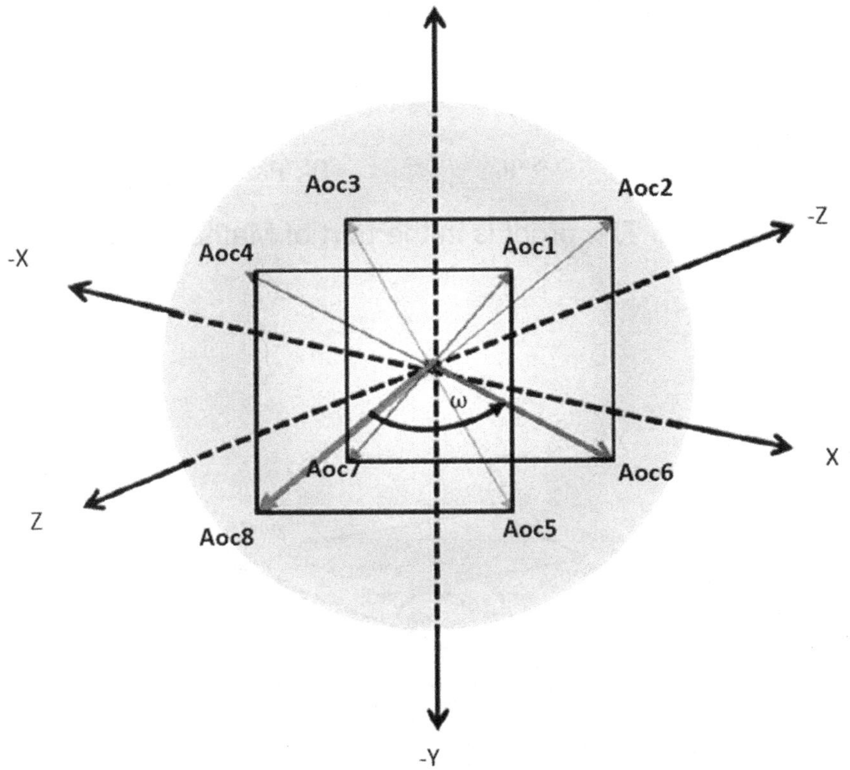

Figure 51: One of the cosine angles Omega made by two resultant directions of the sixth and the eighth parts, Aoc6 and Aoc8, in the three-dimensional body.

This figure shows how cos ω_{68} looks like.

The angle of φ as 70.5287°

There are twelve angles of the cosine angles Phi φ as 70.5287° in the three-dimensional body (Boonsri, 2003). They are as follow:

$$\cos 70.5287^{\circ} \quad = \quad \cos \varphi_{12} \quad = \quad \cos \varphi_{14} \quad = \quad \cos \varphi_{15}$$

$$= \cos\varphi_{23} = \cos\varphi_{26} = \cos\varphi_{34}$$

$$= \cos\varphi_{37} = \cos\varphi_{48} = \cos\varphi_{56}$$

$$= \cos\varphi_{58} = \cos\varphi_{67} = \cos\varphi_{78}$$

See Figure 52. The proof is in the part of Mathematical Proof of chapter 4.

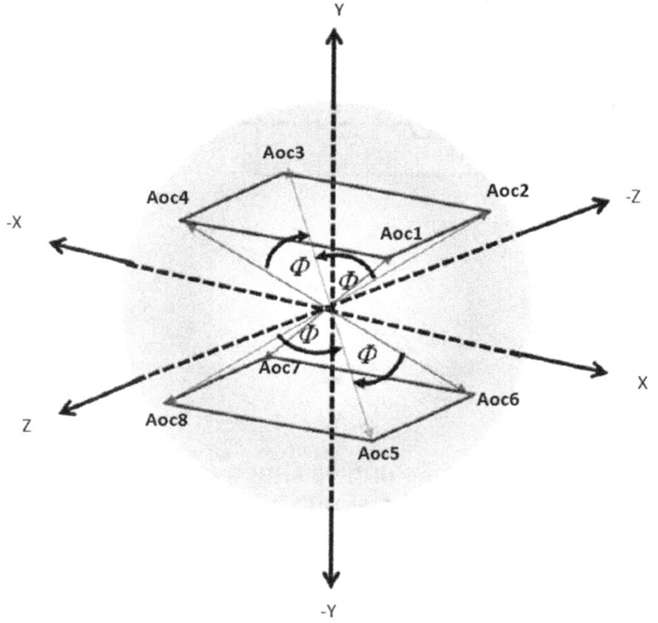

Figure 52: Four of the angles Phi made by two resultant time directions of the upper and the bottom time-fields in three-dimensional body.

The angle of δ as 180°

There are four angles of the cosine Delta δ as $180°$ in the three-dimensional body (Boonsri, 2003). They include:

$$\cos 180^\circ = \cos\delta_{17} = \cos\delta_{28} = \cos\delta_{35} = \cos\delta_{46}$$

See Figure 53. See the proof in the part of mathematical proof of chapter 4.

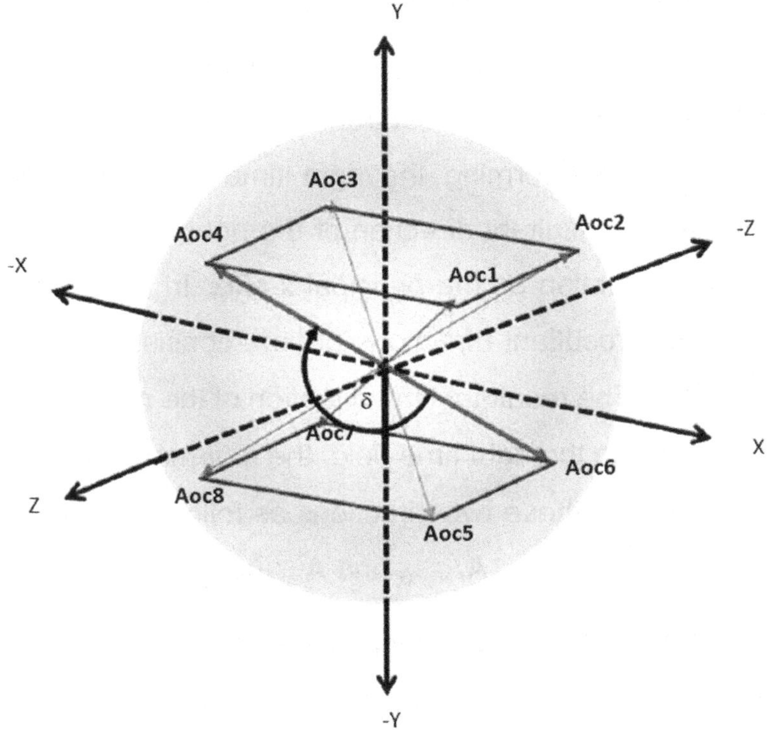

Figure 53: One of four angles made by two resultant directions from different time-fields.

In the above figure, the structure of cos δ_{46} in a 3-D model is shown.

Scientific evidence of existence of the time dimension in a three-dimensional body: the magic angle as 54.7356°

The implication of time at the subatomic level came from finding that time is linked to the magic angle, which physicists know well that the magic angle is in subatomic level. There are

24 different mathematical ways to analyze that how time is involved at the subatomic level (Boonsri, 2016).

The magic angle is created by two resultant directions of time as well. For example, the first direction is one of four resultant directions of time forming the right time-field. The second direction is the resultant direction of the right time-field which is as same direction as the principal x-axis. In the right time-field, the four resultant directions of time consist of $\mathbf{A}_{oc1}$, $\mathbf{A}_{oc2}$, $\mathbf{A}_{oc5}$, and $\mathbf{A}_{oc6}$. The resultant time direction of the right time-field is $\mathbf{A}_{PAright}$ or $\mathbf{A}_{X}$. In the right time-field, the four magic angles are created between these two directions as follow: $\mathbf{A}_{oc1}\text{-}\mathbf{A}_{PAright}$ (or $\mathbf{A}_{oc1}\text{-}\mathbf{A}_{X}$), $\mathbf{A}_{oc2}\text{-}\mathbf{A}_{PAright}$, $\mathbf{A}_{oc5}\text{-}\mathbf{A}_{PAright}$, and $\mathbf{A}_{oc6}\text{-}\mathbf{A}_{PAright}$.

In the left time-field, the four magic angles are created between these two time directions as follow: $\mathbf{A}_{oc3}\text{-}\mathbf{A}_{PAleft}$ (or $\mathbf{A}_{oc3}\text{-}\mathbf{A}_{(-X)}$), $\mathbf{A}_{oc4}\text{-}\mathbf{A}_{PAleft}$, $\mathbf{A}_{oc7}\text{-}\mathbf{A}_{PAleft}$, and $\mathbf{A}_{oc8}\text{-}\mathbf{A}_{PAleft}$.

In the upper time-field, the four magic angles are created between these two time directions as follow: $\mathbf{A}_{oc1}\text{-}\mathbf{A}_{PAupper}$ (or $\mathbf{A}_{oc1}\text{-}\mathbf{A}_{Y}$), $\mathbf{A}_{oc2}\text{-}\mathbf{A}_{PAupper}$, $\mathbf{A}_{oc3}\text{-}\mathbf{A}_{PAupper}$, and $\mathbf{A}_{oc4}\text{-}\mathbf{A}_{PAupper}$.

In the bottom time-field, the four magic angles are created between these two time directions as follow: $\mathbf{A}_{oc5}\text{-}\mathbf{A}_{PAbottom}$ (or $\mathbf{A}_{oc5}\text{-}\mathbf{A}_{(-Y)}$), $\mathbf{A}_{oc6}\text{-}\mathbf{A}_{PAbottom}$, $\mathbf{A}_{oc7}\text{-}\mathbf{A}_{PAbottom}$, and $\mathbf{A}_{oc8}\text{-}\mathbf{A}_{PAbottom}$. See Figure 54.

In the front time-field, the four magic angles are created between these two time directions as follow: $\mathbf{A_{oc1}.A_{PAfront}}$ (or $\mathbf{A_{oc1}.A_{Z}}$), $\mathbf{A_{oc4}.A_{PAfront}}$, $\mathbf{A_{oc5}.A_{PAfront}}$, and $\mathbf{A_{oc8}.A_{PAfront}}$. See Figure 55.

In the back time-field, the four magic angles are created between these two time directions as follow: $\mathbf{A_{oc2}.A_{PAback}}$ (or $\mathbf{A_{oc2}.A_{(-Z)}}$), $\mathbf{A_{oc3}.A_{PAback}}$, $\mathbf{A_{oc6}.A_{PAback}}$, and $\mathbf{A_{oc7}.A_{PAback}}$.

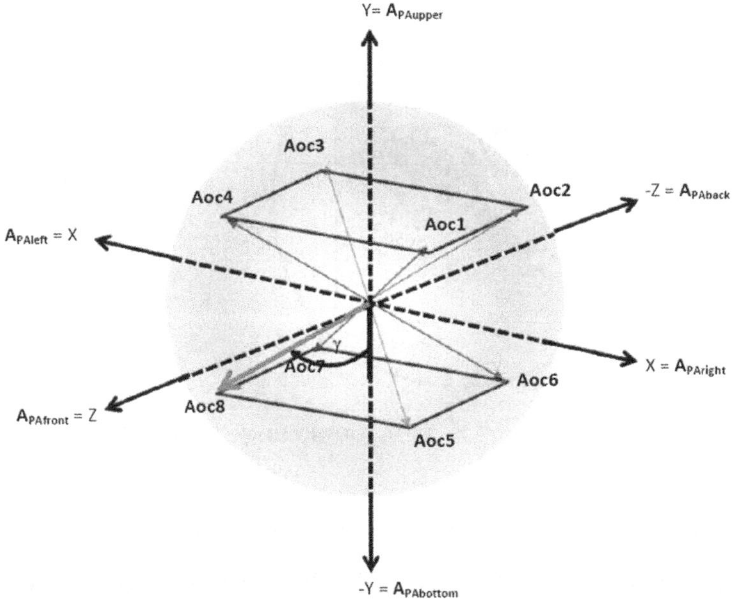

Figure 54: One of twenty-four magic angles Gamma created by two vectors, A_{oc8} and $A_{PAbottom}$ at the bottom time-field.

The figure shows how $\mathbf{Cos\gamma_{8(-Y)}}$ or one of 24 magic angles looks like.

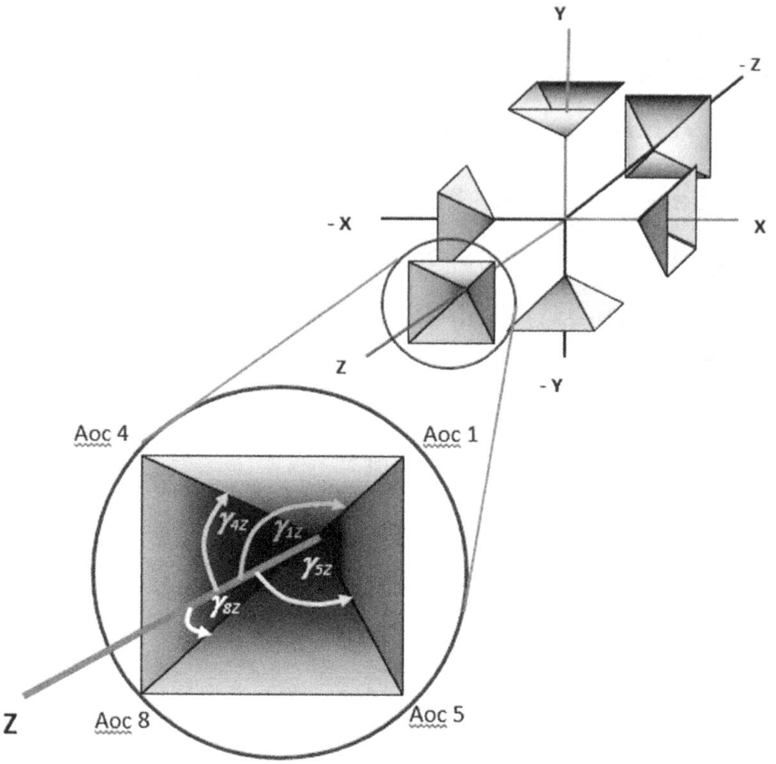

Figure 55: $\gamma_{1z} = \gamma_{4z} = \gamma_{5z} = \gamma_{8z}$ **= magic angle in the front time-field.**

The figure shows the four of 24 magic angles.

The conclusion of the topic is that we have twenty four ways for calculating the magic angle 54.7356°. The values of time in the magic angle show time in the subatomic level. This is the scientific evidence supporting existence of the time dimension in the three-dimensional body.

Magic angle is one of the angles which physicists cannot explain why the angle exists in the subatomic level. Scientist investigated and reported that the angle shows many unexplainable results in subatomic level. This makes it be called "the magic angle." The magic angle is used in studying about the nuclear level such as some characteristics of electron or spectrum. The existence of the magic angle has been reported for a long time, but scientists still do not know that why this not the others.

In the table 2, the vertical direction presents the resultant directions of time of the eight parts in the three-dimensional body, and the horizontal direction presents six resultant directions of six time-fields that are the same directions of three principal axes with positive and negative directions. We found that the ordered triples in the eight parts of the three-dimensional body will show twenty-four ways for getting the magic angle (Boonsri, 2016). See Table 2.

After we had found the time as a vector, the magic angles were found next. This angle, 54.7356°, can be calculated by indirect and direct methods. In addition, time is in subatomic level. In the future, the time dimension may be one of many ways for us to search for the details in many levels of a three-dimensional body such as the tiny world in subatomic level to the big world of celestial level.

Table 2: The existence of the magic angle, 54.7356°, following ordered triples of the three-dimensional body

Resultant Time directions in eight parts	Order triples of the eight parts	Resultant directions of six time-fields in a three-dimensional body especially in subatomic level: $A_{PAright} = A_X = v_x t_x R_x$; $A_{PAleft} = A_{(-X)} = -(v_x t_x R_x)$; $A_{PAupper} = A_Y = v_x t_x R_x$; $A_{PAbottom} = A_{(-Y)} = -(v_y t_y R_y)$; $A_{PAfront} = A_Z = v_z t_z R_z$; $A_{PAback} = A_{(-Z)} = -(v_z t_z R_z)$.					
		X	-X	Y	-Y	Z	-Z
A_{oc1}	(x, y, z)	$\cos\gamma_{1X}$		$\cos\gamma_{1Y}$		$\cos\gamma_{1Z}$	
A_{oc2}	(x, y, -z)	$\cos\gamma_{2X}$		$\cos\gamma_{2Y}$			$\cos\gamma_{2(-Z)}$
A_{oc3}	(-x, y, -z)		$\cos\gamma_{3(-X)}$	$\cos\gamma_{3Y}$			$\cos\gamma_{3(-Z)}$
A_{oc4}	(-x, y, z)		$\cos\gamma_{4(-X)}$	$\cos\gamma_{4Y}$		$\cos\gamma_{4Z}$	
A_{oc5}	(x, -y, z)	$\cos\gamma_{5X}$			$\cos\gamma_{5(-Y)}$	$\cos\gamma_{5Z}$	
A_{oc6}	(x, -y, -z)	$\cos\gamma_{6X}$			$\cos\gamma_{6(-Y)}$		$\cos\gamma_{6(-Z)}$
A_{oc7}	(-x, -y,-z)		$\cos\gamma_{7(-X)}$		$\cos\gamma_{7(-Y)}$		$\cos\gamma_{7(-Z)}$
A_{oc8}	(-x,-y, z)		$\cos\gamma_{8(-X)}$		$\cos\gamma_{8(-Y)}$	$\cos\gamma_{8Z}$	

The detail of the shape of the time-field

The detail of the shape of each time-field in Figure 56 consists of the height, the edge, the half of diagonal, and one of the four sides of the base, these can be presented as follow: (The proof of this section is in the part of Mathematical Proof of chapter 4.)

1. The height of the time-field in the form of symbols of vector is A_1 or A_2 or A_3.

2. The edge of the time-field in the form of symbols of vector is $\sqrt{3}\,A_1$ or $\sqrt{3}\,A_2$ or $\sqrt{3}\,A_3$.

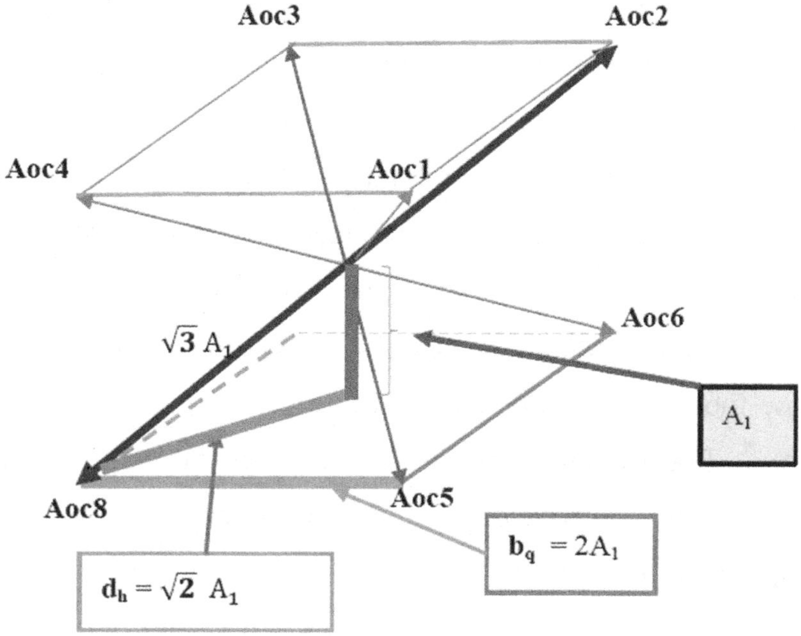

Figure 56: The details of a time-field

The figure shows the shape of time-field to readers. Readers can use these sizes to make a miniature time-field to do any experiments at their houses. As a reader, you can conduct an experiment such as decay of fruit in a time-field (the pyramidal shape) to prove this case.

3. One side of the four sides of the base of the time-field, symbol b_q, in the form of numbers is $2A_1$ or $2A_2$ or $2A_3$.

With (1) to (3), we can use it to make a miniature pyramidal pyramid to do experiments. For example, **first,** make a center of a circle on hard paper. **Second,** if we want the height of the pyramidal shape is 5 inches (or $A_1 = 5$), at the center, make the circle with radius as 8.66 inches (or $\sqrt{3}A_1$ = 1.732 x 5). **Third,** mark a point on the paper to be a starting point (of the base of the pyramidal shape), from the point, draw a straight line with 10 inches long to intercept with the circle (or $2A_1$ = 2 x 5 =10). **Fourth,** make the third step again three times (or 4 x $2A_1$ = 4 x 2 x 5 = 40). **Fifth,** make one and a half inches for form the pyramidal shape. **Sixth,** at the four intercept points, draw four lines to the center. **Seventh,** form the pyramidal shape. The pyramidal shape consists of the edge as 8.66 inches long, one side of the base as 10 inches long, and its height as 5 inches long.

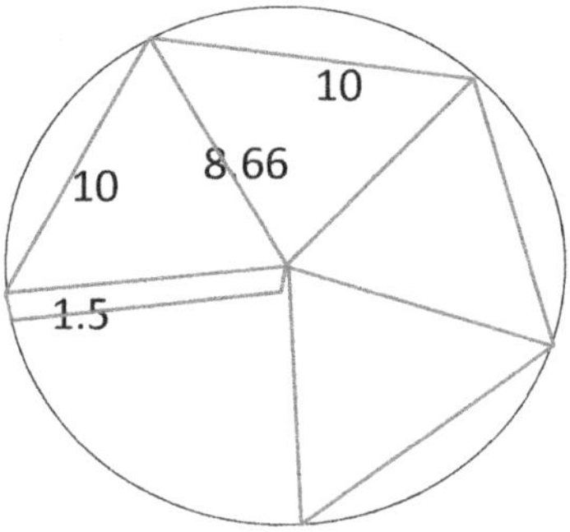

Figure 57: Making a miniature of pyramidal shape

In addition, the shape will have the half of the diagonal at the base as 7.07 inches long (or $\sqrt{2}A_1$ = 1.414 x 5), the diagonal at the base as 14.14 inches long (or $2\sqrt{2}A_1$ = 2 x 1.414 x 5), the four sides of base are 40 inches long (or 8 x A_1 = 8 x 5).

This size of the pyramidal shape was used to do many experiments as follow: (1) to increase and decrease redox chemical reaction including the growth of plants discussed in the fifth chapter, (2) to duplicate my discovery in physics in 1987 discussed in the sixth chapter. The readers can do experiment to see the characteristic of the time-field with monopole.

4. The half of the diagonal at the base of each time-field (d_h), in the form of symbols of vector is $\sqrt{2}\,A_1$ or $\sqrt{2}\,A_2$ or $\sqrt{2}A_3$.

5. The diagonal at the base of each time-field ($2d_h$), in the form of positive numbers, is $2\sqrt{2}\,(A_1)$or $2\sqrt{2}\,(A_2)$ or $2\sqrt{2}\,(A_3)$.

6. The length of the four sides of the base is $4b_q$ = four sides of the base = $8A_1$ = $8A_2$ = $8A_3$.

7. The radius of the perfect sphere, which six time-fields is inside it, is$\sqrt{3}\,(A_1)$ or $\sqrt{3}\,(A_2)$, or$\sqrt{3}\,(A_3)$.

8. The diameter of the perfect sphere, which six time-fields is inside it, is $2\sqrt{3}\,(A_1)$ or $2\sqrt{3}\,(A_2)$ or $2\sqrt{3}\,(A_3)$.

The pyramidal shape as a time-field with monopole

In three-dimensional body, there are six time-fields in minimum. This means that the pyramidal shape is a time-field with the only one pole. It has its own characteristic.

In 1967, one of the three great pyramids at Giza in Egypt called "Chephren" made physicists realize that present physics was not good enough for explaining the pyramidal shape. Dr. Luis Walter Alvarez, the Nobel Prize winner in physics in 1968 and senior high-energy physicist of Lorenz Research Institute ("Luis Walter Alvarez", 2016), who worked on the fundamental particles, planned to search for the secret rooms in Chephren, the second pyramids in Giza, by shooting it with cosmic particles. The huge international project was done by many physicists whose leader was Alvarez.

Alvarez expected that the values of cosmic energy of cosmic particles running through the different kinds of places would give him and his colleague different values. If many valuable things were in unseen chambers, the things would absorb the energy; as the result, the values of energy running through stones would be measured by a Geiger counter more than that running through the secret chamber. The US government cooperated with the Egypt government in searching the secret chamber with the highest technology in physics at that time; the team hoped that the project with the budget of two countries would bring them some valuable answers, which they did not know before.

After having set on all preparation, they got many data from two million times of shooting the particles that were collected in magnetic tapes to analyze by the powerful supercomputers at that time. The people who concerned directly and indirectly the project waited for the good news that they had never heard before.

Unfortunately, the result that the scientists expected to receive from the data was false because the magnetic tapes indicated that the pattern of the high energy cosmic particles cannot be measured. Moreover, they found the other very serious problems too. The patterns of data from the magnetic tapes had been changing all the time that the scientists reprocessed. In addition, the certain data, which the scientists always received, disappeared but no one explained the reason why it happened. They received the useless results ("Luis Walter Alvarez", 2016).

This example indicated that Alvarez, one of the most intelligent scientists of the world, and all the others, became stunned and could not explain what and why really happened. The pyramidal shape was the time-field with the monopole. It forced the team of the Nobel Prize winner to receive the unsolved question. The reporter of The Time magazine wrote with some exciting sentences on the headline that the result of this research was beyond all of present knowledge in physics. In other words, present physics did not reach to the level, which would solve the time-field with the only one pole.

In 1930, one French tourist, Mr. Bovis, made sightseeing around the great pyramid. While he was walking in the King's chamber, he found dead body of a cat and a dog which did not decay; when Bovis asked a guard, he was told that the cat and the dog lost their ways in the chamber and died because of starvation. After returning home, he began to do some experiments of decaying of food in the miniature of a pyramid "he made." He made a record and published the result of his investigating which he found that the food did not decay in the same speed of time as the same food setting outside the pyramid. This characteristic of the time-field with the monopole can be experimented. With my own experience, my friend and I made the hard-paper pyramidal shapes with the height of it as half of the length of the each base. I found the rate of decaying of some fruits inside the pyramidal shape slower than that outside the shape.

Later, Kavel Drbal, electronic Chek engineer, read the Bovis's experiments and did the other experiment. He kept his blunt razor in his pyramid for some time, and he took it out, the razor was able to be used again. And when it became blunt, again he kept it in his pyramid, the razor became sharp again. He did like this for about 200 times. Later, he made his pyramid, which is a sharpener of the razor, and he got patent of it. He called his pyramid "Cheops Pyramid Razor Blade Sharpener." It really worked but no explanation how it worked ("Pyramid power", 2016; "RAZOR SHARPENING PYRAMID by Brazdadj", 2016).

There were some experiments to test the power of the pyramidal shape by taking a container full of water and putting it inside the cell of pyramidal shape for a week, after that, the water is used to water the plants; as a result, the growth rate of the plants watered with the water is faster than that those watered by normal water.

Summary of the chapter

After having found the time as a vector, we used the eight resultant directions of time of the eight parts in the three-dimensional body as the data for searching the geometry of the time dimension in this chapter. The geometry of the time dimension consists of three main parts; six time-fields, the four important angles, and the shape of the time-field.

At the first part, having divided the three-dimensional body into eight parts, we found the eight resultant directions of time in the eight parts. Sometimes, they were called by me "eight octodrants." The eight resultant directions of time can be grouped into six time-fields that are the minimum of time-fields in the three-dimensional body because there have been N directions of time in a three-dimensional body in maximum. Each field has its own direction. We can group these six time-fields into three groups. All of them have the same magnitude but different directions. Three pairs of them can be grouped as follow: the right and the left time-fields, the top and the bottom ones, the front and the back ones.

In case of minimum, the resultant directions of six time-fields show the characteristic of a vector not a scalar. Each time-field has its own characteristic that is different from each other. The left, the bottom, and the back time-fields are in the opposite sides of the right, the upper, and the front ones respectively. The shape of six time-fields in a three-dimensional body looks like six pyramidal shapes with only one vertex. All of six time-fields have their own resultant directions.

The four resultant directions of time of the four parts with numbers 1, 2, 5, and 6 in the three-dimensional body form the right time-field. Moreover, the four resultant time directions in the four parts with numbers 3, 4, 7, and 8 form the left time-field. The resultant directions of the right and the left time-field are $4A_1\mathbf{i}$ and $-4A_1\mathbf{i}$ respectively. Both of them are in the opposite direction, but same magnitude.

Similarly, the four resultant time directions of the parts numbers 1, 2, 3, and 4 form the upper time-field. The four resultant time directions of the parts numbers 5, 6, 7, and 8 form the bottom time-field. In addition, the resultant directions of the upper and the bottom time-field are $4A_2\mathbf{j}$ and $-4A_2\mathbf{j}$ respectively. Both of them are in the opposite direction, but they have the same magnitude as well.

In addition, the four resultant time directions of the parts numbers 1, 4, 5, and 8 form the front time-field. The four resultant time directions of the parts numbers 2, 3, 6, and 7 form the back time-field. The resultant directions of the front and the back time-fields are $4A_3\mathbf{k}$ and $-4A_3\mathbf{k}$ of the three-dimensional body. Both of them are in the opposite direction, but they have the same magnitude. All six time-fields have only one vertex. They show the characteristic of a vector and are calculated by the mathematical vector.

In the second part, the four kinds of angles are $\omega, \varphi, \delta,$ and Magic γ, the values of which are approximately 109.4712°, 70.5286°, 180°, and 54.7356° respectively. For getting these angles, we have 12, 12, 4, and 24 mathematical ways respectively. Magic angle γ is the space between one of the eight resultant time directions of the eight parts and one of six resultant directions of six time-fields presented in Table 2. It is the scientific evidence of existence of the time dimension in three-dimensional body especially in the subatomic level.

In the third part, the six time-fields in three-dimensional body are six pyramidal shapes with a vertex. The shape of the time-field has the ratio between its height and a side of square base as 1: 2 that was used to do many experiments presented in the fifth and the sixth chapters.

With following these vector time-fields, the one of the secrets of the universe is the shape of the pyramid. Its shape directly concerns space and time dimensions in a three-dimensional body. The three-dimensional bodies can be things such as the small in subatomic level and the big in the heaven. This may mean that the resultant directions of the six time-fields inside all of particles, atoms, molecules, cells of creatures, plants, and any objects in our universe have been done.

In the future, the geometry of the time dimension will be one of the parts of which present physicists have to discuss more about. One of the new characteristics of time dimension is the directions of time-fields.

The result of interaction of these six time-fields in a three-dimensional body A and another body B was discussed in the sixth chapter in this book.

Before the time-field was found, the characteristic of symmetry had appeared many times. When three principal axes divide the whole space of the three-dimensional body into eight parts, which each part is the same but different ordered triple, the symmetry inside the three-dimensional body shows itself in this level. The eight resultant directions of time have the same magnitude but different directions; the symmetry appears in this level. Those six time-fields have the same magnitude but different directions; the symmetry still appears in this level again.

In addition, the magic angle is also the scientific evidence supporting existence of the symmetry in the subatomic level. Everything in the universe is moving. All of them are dynamics. Everything in the universe is governed by time. Time is in everything in the universe including force.

Chapter 5
New Discoveries from Einstein's Work

We cannot get to where we dream of being tomorrow unless we change our thinking today

Albert Einstein

In 1987, my friend named Apichai Sivapraphagorn and I did some experiment concerning decaying rate of fruits in pyramidal shape. After geometry of time dimension had been presented in national scientific conference in Thailand, my friend started to deeply study about the shape in chemistry. During 2004 to 2006, Sivapraphagorn, an Associate Professor in Chemistry of Ubonratchathani University located at Ubon Ratchathani Province at the northeastern Thailand, did his researches to use the time-field with monopole for accelerating or decreasing values of time in the redox reaction (Sivapraphagorn, 2006). See Figure 58.

Time-field in Chemistry: decreasing and increasing time of redox reaction and plants

Until now, chemists have studied chemical reaction and concluded that chemical reaction is a process, by which chemical substances have changed to create new substances. The new substance called product carries different chemical characteristics from its original state. The chemical reaction needs to have at least two substances called reactants reacting together. However, the chemical characteristics of some product may change so little from the reactants, some may change completely.

Figure 58: Apichai Sivapraphagorn, Associate Professor

In the past, the study of the chemical reaction emphasized only movement of electrons, which created some chemical bonding. With the only traditional definition of chemical reaction, chemical reaction consists of two kinds of reactions; Redox reaction and acid-base reaction. Redox reaction concerns the transfer of electron charges, and the acid-base reaction concerns the movement of the pair of electrons.

Redox reaction is divided into two parts; (1) reduction is a reaction that donates electrons, and (2) oxidation is a reaction that gains electrons. In case of creating a product, if we synthesize chemical substances, the chemical reaction will occur. In biochemical field, if chemical reactions occur continuously; as the result, the metabolic pathway is created. The synthesized product is not created in cells in only a moment because the energy in cells is not enough. The redox reaction is in all cells of living things; animals and plants.

If the pyramidal shape time-field affected redox reaction, then he would control metabolism in any cells in the future (Sivapraphagorn, 2006). For example, increasing the redox reaction to accelerate metabolism path of plant growing will solve the food problems and heal wounds from some injuries of people and animals. In contrast, decreasing the redox reaction to slow metabolism path of aging down will decrease cancer cells, HIV viruses, and decaying rate of food.

In 2011, Sivapraphagorn and his team that consisted of Supon Samran, a physicist, Aranya Pimmongkol and Warinee Parasarn, two biologists working in the Faculty of Science, Ubonratchathani University, prepared his experiment concerning the time-fields in three steps. See Figure 59. **First,** he made the activating process of the time-field by the arrangement of electrolytic capacitors along the edges of the time-field under a magnetic field.

Second, after the activation process had been completed, he investigated the rate of chemical reaction under a time-field which compared with the control which is called “the blank.” **Third**, he investigated the growth of the tissue culture plants under a time-field which compared with the blank (Sivapraphagorn, Samran, Pimmongkol, & Parasarn, 2011).

For activating process, he made a hard paper time-field with the height of 150 mm and one side of the base of 300 mm long at the beginning. Four triangular metal sheets were attached inside the four sides of the pyramidal shape with a one-centimeter gap between them. See Figure 60.

Figure 59: A physicist, Supon Samran; two biologists, Aranya Pimmongkol and Warinee Parasarn respectively.

Electrolytic capacitors of 16 Voltages, 1,000 microfarads were placed along the four sloping edges. He placed a positive pole of each capacitor on the upper direction of each declining sloping edge. See Figure 60. He also bent the two poles of all electrolytic capacitors to contract the metal sheets. Only the positive or negative poles of electrolytic capacitors contracted each metal sheet.

The opposite sides of the metal sheets that will be positive or negative depend on the ways of pole arrangements. The structure was placed on a ceramic plate with a cylindrical magnet of a speaker under the center of the plate. Sivapraphagorn calls it "a positive time-field structure." See Figure 61. To made "negative time-field structure" with hard paper, he placed a negative pole of each capacitor on the upper direction of each declining sloping edge. The difference in this structure from the positive time-field was the opposite arrangement of the electrolytic capacitors. See Figure 62.

Both positive and negative time-field structures had been activated before experiments began. The time-fields were placed in the experiment area without any disturbance for at least 10 days. This period must be extended if the relative humidity is higher than 75%.

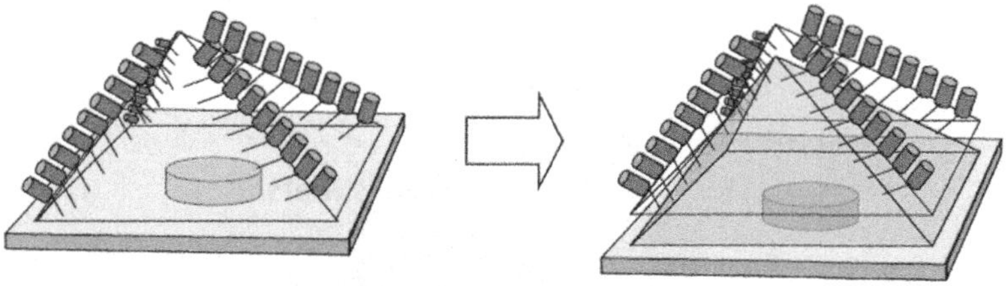

Figure 60: Activating process of the hard paper time-field was done. Metal sheets with a one-centimeter gap were placed. A ceramic plate with magnet was set under the plate.

The figure shows how to activate time-field.

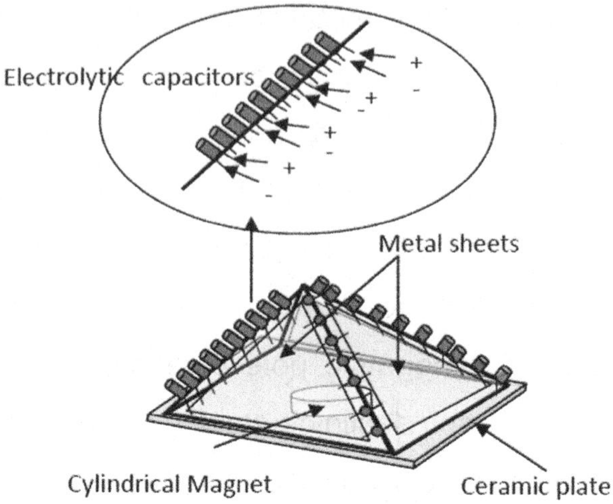

Figure 61: A positive pole of each capacitor set along an edge is on the upper direction of each edge, this form is called "a positive time-field structure."

The figure shows how to make a positive time-field to do experiment.

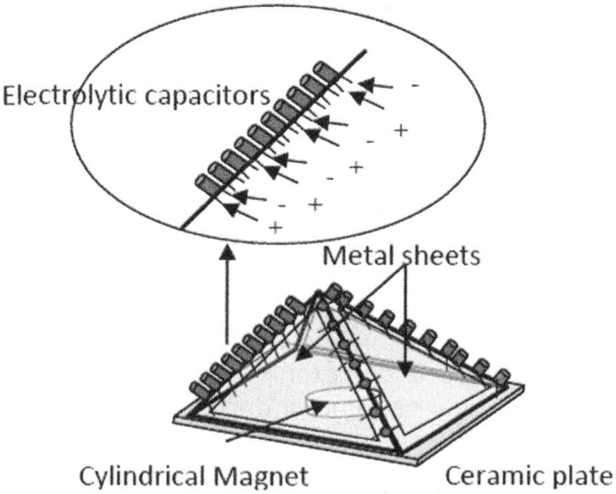

Figure 62: A negative pole of each capacitor set along an edge is on the upper direction of each edge, this form is called "a negative time-field structure."

The figure shows how to make a negative time-field. The form is called "a negative time-field structure."

Upon activation and completion of the process, the experiments to control Redox reaction by using the time-field with the only one pole must be done only at the activating location. A blank, an inactivated time-field structure, was also made of paper, similar to the activated time-field but without any electrolytic capacitors. The structure was placed on a ceramic plate with a cylindrical magnet at the center of the plate as well.

The activating process and the setting of the blank were done at the same time. The blank structure was placed within 300 millimeters from the activated time-field for comparing reactions between the blank and the activated time-field.

He used an iodine clock reaction to show chemical reaction between the time-field and the blank. The different time of the two iodine clock reaction is measured in seconds. When the reaction is completed, the color of the reactants changes from colorless to blue products immediately. The rate of reaction is the clearance of the reaction, which everyone can see.

A stock solution of iodine clock reaction was prepared by mixing 200 milliliters of Sodium thiosulfate ($Na_2S_2O_3$) solution which has a concentration of 0.2 moles $Na_2S_2O_3$ per liter of solution (0.2 M), 80 milliliters of acetate buffer (pH 4.74), 80 milliliters of Potassium Iodide (KI), which has a concentration of 0.3 moles KI per liter of solution (0.3 M), and 10 milliliters of starch solution. A solution of Hydrogen peroxide (H_2O_2), which has a concentration of 0.1 moles H_2O_2 per liter (0.1 M), was used for reacting to the stock solution. After the completed reaction, the solution turns blue.

He divided the two homogeneous solutions into two beakers to investigate the rate of reaction in the time-field and the blank. The homogeneous solution was prepared by mixing 10 milliliters of 0.1 M H_2O_2 with 25 milliliters of stock solution for 24 times transfer between two beakers.

When the two solutions reacted, he commenced the time by a watch. After the homogenizing process was completed, equal amounts of homogeneous solution were transferred into the two beakers. This step was done carefully with checking for the homogeneity of the reaction to be completed within 7-9 minutes.

Because of the room temperature affecting the result of the experiment, he sometimes adjusted the concentration of Hydrogen peroxide, H_2O_2, so that the reaction time would be in the range of 7-9 minutes. The two beakers of homogenous solution were placed in the blank structure and in the positive time-field structure.

Before the reaction was completed, the two structures, the positive time-field and the blank structures, had been opened, and time was stopped when the solution turned blue. The time difference was recorded as the effect of a positive time-field. The reaction tests were performed every week for eight weeks. The results of the three replications of the weekly test were averaged and recorded in Table 3.

Table 3: The different times between two structures

Days	Time (seconds)		
	Time-field (T)	Blank (B)	Difference (T< B)
0	0	0	0
7	8:46.56	8:49.70	3.2
14	9:39.82	9:43.92	4.1
21	8:20.85	8:24.25	3.4
28	7:54.94	7:58.04	3.1
35	8:35.68	8:37.98	2.3
42	9:46.73	9:49.73	3.0
49	9:51.76	9:56.44	4.7
56	8:17.81	8:21	4.3

The result of the positive time-field started after 10 days of activation had done. The experiment was done every week. He recorded two iodine clock reactions between the time-field and the blank. The different times of both were calculated in Table 3 and plotted in Figure 63. The different times are between 2.3 and 4.7 seconds with an average of 3.5 seconds in 7-9 minutes reaction time. He found also fluctuation of the different times, which may have been caused by unstable humidity during the period of 56 days.

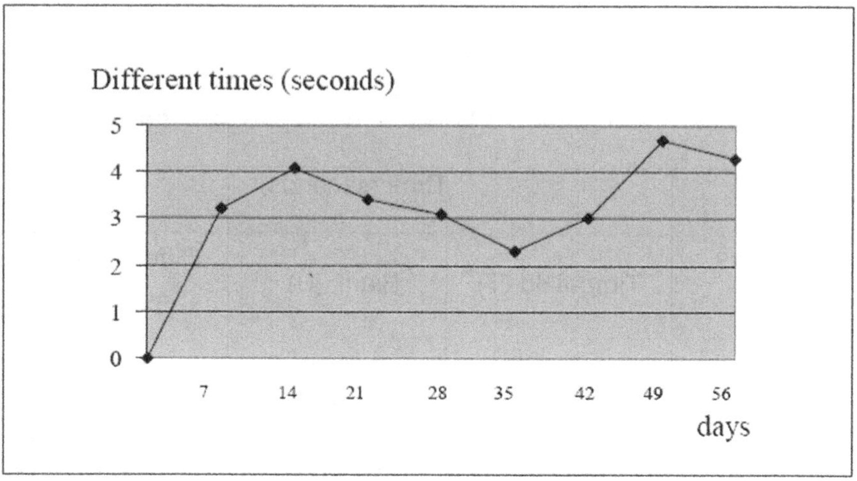

Figure 63: The different times of reactions of two structures were plotted.

The figure shows the result of the experiment. Similarly, the experiment of a negative time-field structure was done. The results of the three replications of the test were averaged and recorded in Table 4.

The experiment with the negative time-field started after 54 days of activation had done. Activation time in negative case is more than the positive case. The range of the different times is -1.3 to -4.7 seconds, with an average of -2.3 seconds. See Figure 64.

Table 4: The different times of two structures

Days	Time (seconds)		
	Time-field (T)	Blank (B)	Difference (T > B)
0	0	0	0
55	8:45.78	8:44.38	-1.4
84	8:21.62	8:16.92	-4.7
97	9:54.36	9:52.36	-2.0
99	8:39.79	8:38.39	-1.4
101	7:26.48	7:24.48	-2.0
116	7:57.82	7:55.42	-2.4
117	9:54.61	9:51.71	-2.9
118	9:43.83	9:42.13	-1.7
119	9:32.57	9:30.77	-1.8

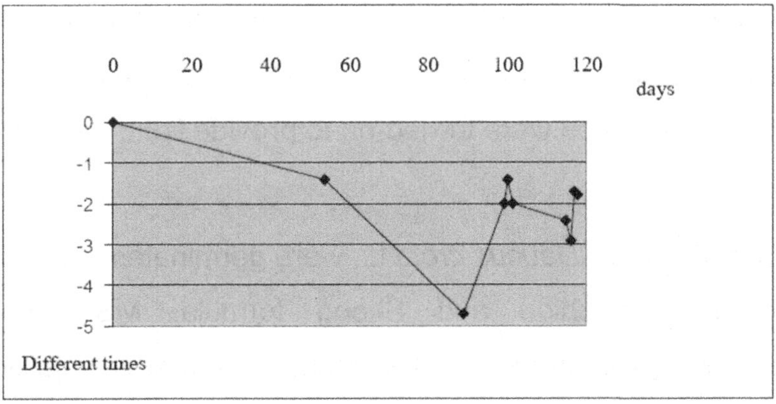

Figure 64: The different times of reactions of two structures were recorded.

In short, the result of the positive time-field structure indicates that the redox reaction can be accelerated. In other words, the acceleration of metabolism path of plant growing is positive. In contrast, the result of the negative time-field structure reveals the positive possibility of decreasing of the redox reaction to slow metabolism path of cells down.

These experiments made him realize that the Arrhenius equation, which chemists always used to explain chemical reaction, has not completed yet.

After the results of both positive and negative time-field structures had been found, Sivapraphagorn and his colleagues did their next experiments with tissue culture plants.

Before the experiment started, the time-field and the blank had been activated for a month in a plant tissue culture room. Fluorescent lamps were turned on to provide light. See Figure 65.

Seeds of Carrot, *Daucus crota* L., were germinated in medium following Murashige and Skoog formula, MS, without hormones for eight weeks. The stems of the seedlings were cut to 1 cm long and then placed on MS medium containing 2 milligrams L-1 2, 4-D for callus initiation. After eight weeks, four calluses (0.5 cm in diameter) were transferred to a bottle of fresh callus induction medium.

Figure 65: Nine bottles of plant material in the time-field (front) and other nine bottles in a blank structure (rear).

The above figure is setup for the experiment. In the experiment, these two directions of change are evidence for the possibility of supporting the existence of time vectors that can be composed to become a time-field; they indicated that a practical proof of the time dimension as a vector is possible after activation.

They used African violet (*Saintpaulia* sp.) plantlets regenerated from leaf disc culture in MS containing 0.2 milligrams L-1 NAA; and 0.1 mg L-1 BA were used as other plant material. A plantlet was transferred to a bottle containing fresh medium. (NAA is 2-(1-napthyl) acetic acid, BA is 6-benzyladenine) (Murashige & Skoog, 1962).

After two months of testing, the fresh weight of both the carrot callus and the African violet plantlets in the time-field increased by 0.1876 and 1.2284 g respectively comparing with the blank. The average percentages of increased fresh weight of the carrot callus and African violet plantlets were 10.11% and 11.44%, respectively.

In general, chemists use Arrhenius equation to explain the change of the rate of a chemical reaction. In the Arrhenius equation, $k = Ae^{-Ea/(RT)}$, only the frequency factor causes the change of the rate of a chemical reaction. The increase in positive time-field and the decrease in the negative time-field affect the frequency factor in the Arrhenius equation too.

In addition, the time-field itself causes time dilations (both positive and negative), which are related to the nature of the opposite properties. Not only ordinary chemical reactions were affected, but also chemical reactions in living organisms.

These scientists did their experiments concerning the time-field. The valuable results are as follow; **first,** they can activate the time-field. **Second,** if they want to accelerate the redox reaction, they use the positive time-field structure. **Third,** if they want to decrease the time of the redox reaction, they use the negative time-field structure. **Fourth,** for accelerating the redox reaction of the plant tissue culture, they use the positive time-field structure. **Fifth,** for decreasing of time of the redox reaction of the plant tissue culture, they use the negative time-field structure. **Sixth,** accelerating and decreasing of time of redox reaction show that the Arrhenius equation, which is used to explain the rate of a chemical reaction with the frequency factor only, cannot cover the process concerning the time-field. In other words, for completeness of the Arrhenius equation, chemists have to add time in the time-field as new factor in the equation. **Seventh,** the time-field can be controlled to show its direction by using electrolytic capacitors attaching at the steep four edges.

In addition, Sivapraphagorn, Samran, and I did another experiment to use the vector time-field to produce free electrical energy for solving energy problem in the near future. This experiment that was presented in National conference of science and technology of Thailand in 2012 and published in Journal of Science and Technology, Ubon Ratchathani University in 2013 was called "Feasibility of an Electromagnetic Cell." This research aimed to create a renewable energy cell independent of solar energy. The magnetic field was selected as a source of renewable energy, electricity by capacitors. A distinctive structural arrangement of electrolytic capacitors was made that involved pooling and tightening together the positive ends of 28 electrolytic capacitors at 54.7 degrees (Magic angle). The negative ends were stretched and connected together with a circle wire. The 28 electrolytic capacitors were connected parallel together to 0 degrees as the blank structure. Both structures were placed in the magnetic field. The wheel structure was charged immediately. An amount of electric charge was measured as electrical potential up to 1.20×10^{-1} V in 20 days while the blank potential was 0.25×10^{-2} V in 20 days (Sivapraphagorn, Boonsri, & Samran, 2014). This paper was not translated in English yet and was not discussed in this chapter.

Summary of the chapter

The activating process of the time dimension by the arrangement of electrolytic capacitors along the resultant time vectors under a magnetic field was investigated. After completion of the activation process, the investigation aimed to find out influences on the rate of chemical reaction and growth of tissue culture plants under a time-field compared with the blank. The experiments showed an average increased rate and an average decreased rate of the redox reactions in a positive time-field and a negative time-field, respectively. Furthermore, the positive time-field resulted in an average of 10.11% and 11.44% increased growth rate of callus culture plants of carrot (Daucus carota subsp. sativus) and African violet (Saintpaulia ionantha) respectively.

The increase in positive time-field and the decrease in the negative time-field affect the frequency factor in the Arrhenius equation which directly concerns the rate of a chemical reaction. Evidence supporting the time-field as a vector can be proved by changing the ways of setting legs of capacitors along the four edges of the hard paper time-field. The two directions of the change indicated that a practical proof has been formed possibly after activation.

In addition, both positive and negative time-field themselves cause time dilations, which are related to the nature of the opposite properties. Ordinary chemical reactions and chemical reactions in living organisms were affected.

Sivapraphagorn was the first one who discovered the process of activation. He and his team were the first team who tested the question about time vector by doing scientific experiment. They helped us to understand the relation among physics, chemistry, and biology in one coherent picture.

The experiment in chemistry can confirm that the time-field has its own pole because the changes of the legs of electrolyte capacitors from electrical positive pole up to electrical positive pole down can be stimulated to perform the different result directions of the plant growing. The existence of the directions of the poles of the time-field can be shown differently. In other words, the relativity between the electrical poles of the capacitor legs and the directions of the edges of the pyramidal shape time-field will have been performing all the time if we set this condition (Sivapraphagorn, Samran, Pimmongkol, & Parasarn, 2011).

Chapter 6
Unsolved Mysteries in Science Solved!

To arrive at the simplest truth requires years of contemplation.

Isaac Newton

Before I started to study about time, in 1987, I discovered the forming of external magnetic poles. At that time I did not know what I found. It was a very strange phenomenon directly concerning time and magnetic field for me. And I directly pushed me to spend my time more than 29 years to get the answer.

Untold discovery in forming of magnetic poles

In 1987, I found there was something wrong in theoretical economics about time after I had read "the Poverty of Economics" published in Atlantic Monthly written by Robert Kuttner and had tested about time in the demand and supply curves by using real world data (""The Poverty of Economics" by Robert Kuttner, The Atlantic Monthly, February 1985", 2016).

In March 1987, I visited my close friend, Mr. Sivapraphagorn, at the Chaiyapruk Flat in Khon Kaen University campus, Khon Kaen Province, Thailand in the room owned at that time by

Sivapraphagorn. Over there, I began to create my first assumption of time especially directions of time.

After having first conclusion about time, my friend and I began to be interested in the shape of the pyramid which the ratio between the height and the length of a side of base of the pyramid is 1:2. The shape was different from the three great pyramids in Egypt, and we made two small pyramids with hard paper for some experiments.

One of the results of the power of pyramid was the decay rate of food. First of all, after the hard paper pyramidal shape with the new size had been made, we had to search for two main directions as north and south at the floor of the room first to put the shape on the floor. By putting a blade on the surface water in a bowl, we found the north and the south easily. The pyramidal shape had been set by fixing the two sides of the base under 90 degrees with the directions. After that, we put some tomatoes inside the pyramidal shape, sealed four sides of its base with glue tape to protect the air from outside, and put some tomatoes on the floor outside the pyramid to compare the decay rate. After three days, we opened the pyramid; we saw that the decay rate of the fruits inside the pyramid had been slower than that of the fruits outside. In addition, we did experiments again and again by changing the types of fruits, and before we put fruits in the pyramidal shape we had searched for the north and the south. We got the same result. (Readers can do the experiment by themselves.)

My theory about the difference of the rate of fruit decay between inside and outside pyramidal shape was about direct electrical current, so we attached some electrolyte capacitors with different sizes such as ten microfarad and a hundred microfarad and else at the four edges of the pyramidal shape without arrangement of the capacitors, and also put some fruits inside the pyramidal shape too. The result of the rate of the decay of fruits was the same. Nothing was different, so we did not pay any attention about the rate of decay again and put the shape with capacitors under a side-square table in the room.

In April 1987, before I discovered in physics, the lines of magnetic field at the floor of the room were normal. The lines always pointed to the north and the south that it is normal in everywhere. Nearly the end of the month, while my friend was working in his office, I woke up and put a blade on the surface of water in a half full bowl of water to see the north and the south. The blade did not point to the north and the south again. It pointed to the east and the west. I thought the wave of water in the bowl as the cause of the abnormal direction, so I used my finger to push the blade to the north and the south again and again. The blade pushed against my finger as if the east and the west were the north and the south of the earth. I concluded immediately that something strange about the direction of the magnetic field at the floor in the room had already taken place.

After my friend came back from his work in the evening and prepared himself to go out for field researching outside the campus, I told him to have a look at the strange unexplainable phenomena about the direction of the head and the tail of the blade, but he did not pay much attention to it because he rushed to go outside. My friends teaching in the faculties of science, agriculture, and engineering could not tell me what it was and how to control the phenomena occurring again.

I tried to find out the directions of the magnetic field at the floor in the room by borrowing a compass from my friend, Dr.Boonmee Siri, See Figure 66. I put the compass to plot the head and the tail at first step, lift and put the head and the tail at the floor again and again. See Figure 67. After that, I found three kinds of lines of the magnetic field on the floor in the room: (1) the areas that the north pole of the compass were pointing to them at that moment, (2) the areas that the south pole of the compass were pointing to them, (3) the areas that both the north and the south poles of the compass were not pointing to them. See Figures 68 to 70.

Figure 66: Dr. Boonmee Siri, Associate Professor

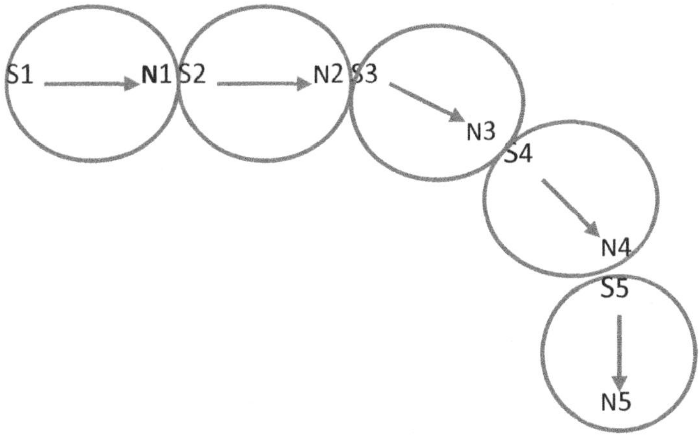

Figure 67: Using a compass to draw a line of magnetic field in the room I discovered.

The figure shows how to plot all magnetic lines on the kite paper to get the shape of three kinds of poles; (1) positive poles, (2) negative poles, and (3) the neutral poles.

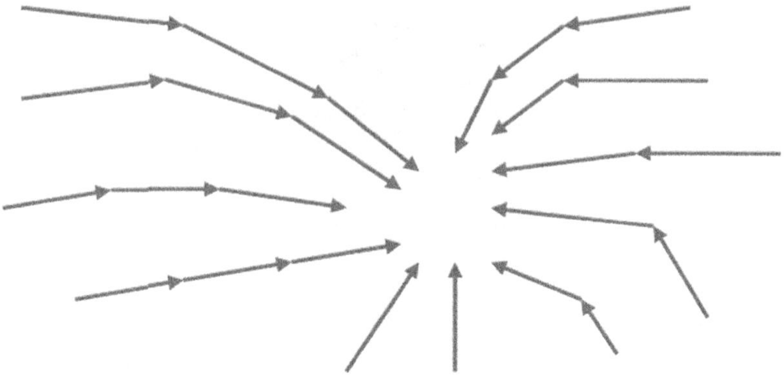

Figure 68: The areas that the north pole of the compass was pointing to them (negative areas)

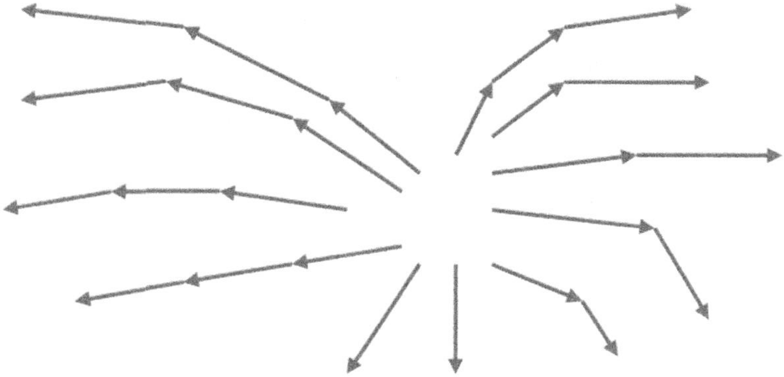

Figure 69: The areas that the south pole of the compass was pointing to them (positive areas)

In figure 68, the shape of negative poles, figure 69 shows the shape of positive poles, and in figure 70, there is the shape of neutral poles.

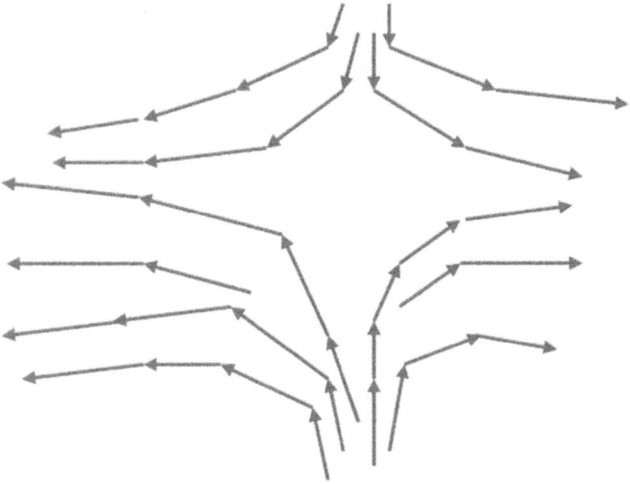

Figure 70: The areas that both the north and the south poles of the compass were not pointing to them (neutral areas

After two days, my friend came back and saw the directions of the lines of magnetic field, he said he did not know the reason why these strange phenomena occurred, and he could not accept the phenomena as scientific method. I debated that although we did not know the reason why it happened, but many discoveries in science in the past happened accidentally. Moreover, there were two main ways for studying science; inductive and deductive methods. With deductive method, we could find out an answer of a new problem from the theory we have known for a long time because the answers of the new problem like a subset in the theory. In other words, we found the answer that could be explained by the theory. The deductive method occurs after a theory has already been established. In contrast, scientists had used inductive method to accumulate many data, before the theory was created. For example, Einstein's special relativity theory and Planck's equation were created by inductive method. At that time, the best way we could do was to collect all the data we found. My friend accepted this view and we plotted the lines of magnetic field on the floor in the kite paper. The figure 71 showed the direction of the lines of magnetic field on the floor at the area.

After that, having changed the area of another experiment, we put two pyramidal shapes with capacitors behind the room. About ten days, we found the three kinds of magnetic lines in new area, and plotted it in some of kite paper again. See Figure 72. We had to stop searching because my friend went to study abroad, and I returned to Bangkok with my puzzle (Boonsri, 2012).

One year later, my friend came back Thailand and we recorded what we found and plotted in standard American VHS tape. See Figures 71 and 72.

While we are recording our tape to show magnetic lines at the area behind the room, our friend, Somsak Pitaksanurat who worked as a lecturer in Faculty of Public Health, visited us; we recorded his visiting in the tape too. Right now, our friend has been working as assistant professor at Faculty of Public Health at Khon Kean University.

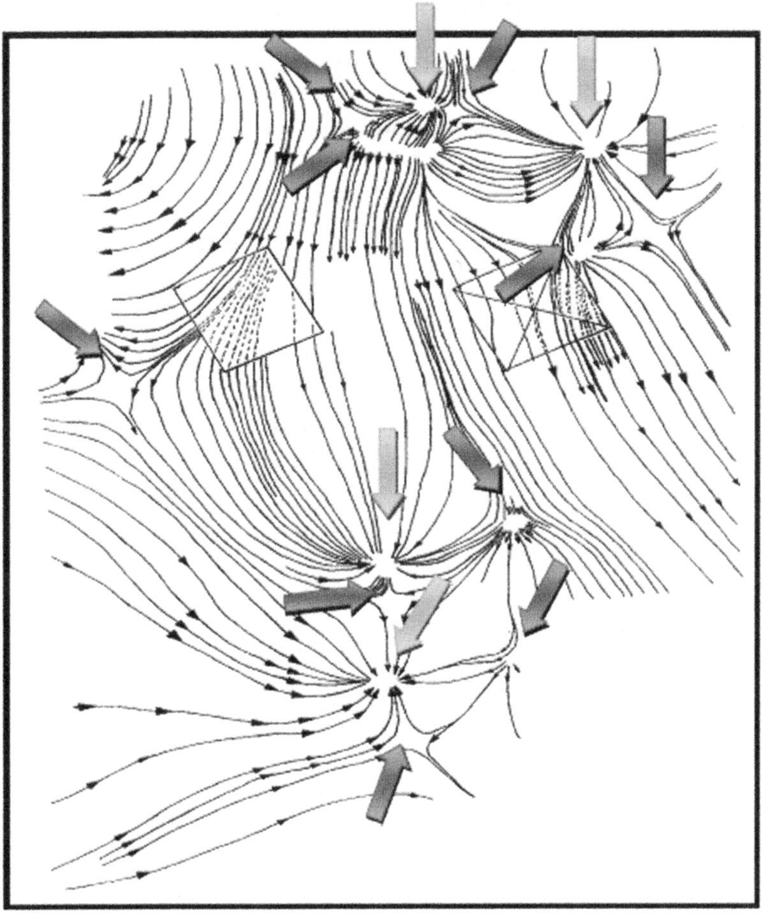

Figure 71: Three kinds of the areas of magnetic field in the room: negative areas, positive areas, and neutral areas

The figure shows what my discovery in physics in 1987 looked like. All magnetic lines on the kite paper were used to get the shape of three kinds of poles; (1) negative poles implies that the heads of arrows were pointing to them; (2) positive poles are the tails of arrows that point to them; and (3) the neutral poles indicate that there are no tails and heads of arrows that point to them. The dimensions of the area are about 2.54 meters wide and 3.13 meters long.

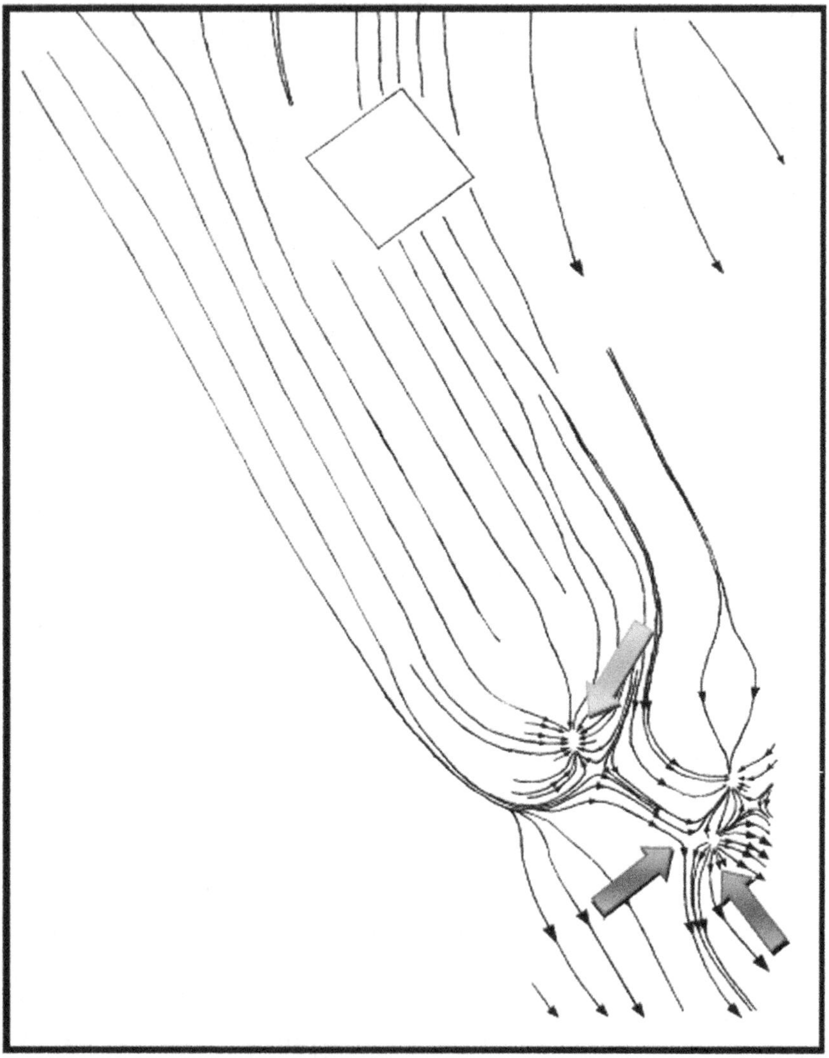

Figure 72: The magnetic lines at another area behind my friend's room.

In that time, I noticed that a little evidence showed me that it is about time and the pyramidal shape. Having paid attention on researching on the time dimension, I found many characteristics of time discussed in the second chapter to the fifth one. Finally, the answers of my questions were that I discovered the formation of external magnetic poles. In addition, the secret of new characteristic of gravity was one of the by-products from studying about time as well.

The two bodies and interactions

What happens when two three-dimensional bodies are moving near each other? A 3-D body has its own six resultant directions of six time-fields. If two 3-D bodies are moving near each other, then 12 resultant directions of 12 time-fields have to be considered. See Figure 73.

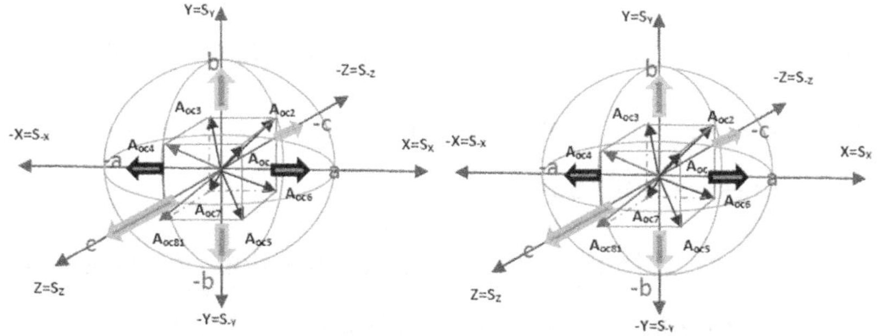

Figure 73: The formation of poles between two 3-D bodies rotating.

The figures above represent the formation of three kinds of magnetic poles that comes from the interaction between two

3-D bodies. One of the 3-D bodies has six resultant directions of six time-fields, so when we consider the interaction of two 3-D bodies, there are 12 resultant directions of 12 time-fields that are used to find the resultant directions. The result of such calculations is shown in Table 10 where three key points are indicated; (1) positive areas, (2) negative areas, and (3) the neutral areas.

When interaction between 12 resultant directions from 12 time-fields occurs, we always see the result of the interaction of two bodies in nature such as two sources of wave are being done.

The six resultant directions of six time-fields of a 3-D body were presented in the topic of the resultant directions of six time-fields in the fourth chapter. The interaction of the two 3-D bodies can be explained by using 12 resultant directions of 12 time-fields of two 3-D bodies. The result of the interaction shows three main different areas taking place as follow:

Define the six resultant directions of the six time-fields in a three-dimensional body as follows:

1) $4A_1i$ = "a" that is the resultant direction of the right time-field in a three-dimensional body.

2) $-4A_1i$ = "-a" that is the resultant direction of the left time-field in a three-dimensional body.

3) $4A_2j$ = "b" that is the resultant direction of the upper time-field in a three-dimensional body.

4) $-4A_2j$ = "-b" that is the resultant direction of the bottom time-field in a three-dimensional body.

5) $4A_3k$ = "c" that is the resultant direction of the front time-field in a three-dimensional body.

6) -4A3k = "-c" that is the resultant direction of the back time-field in a three-dimensional body.

Table 5: The result of interaction between two 3-D bodies

The Second body	The six resultant directions of six time-field of the first body					
	-a	a	-b	b	-c	c
a	-a+a = 0 Neutral	a+a= 2a positive	-b+a mixed I	b+a mixed II	-c+a mixed IV	c+a mixed III
-a	-a+(-a) = -2a negative	a +(-a) = 0 Neutral	-b+(-a) mixed II	b+(-a) mixed I	-c+(-a) = -c-a mixed III	c+(-a) = c-a mixed IV
b	-a+b = b+(-a) mixed I	a+b = b+a mixed II	-b + b = 0 Neutral	b+b = 2b positive	*-c+b* mixed VI	c+b mixed V
-b	-a+(-b) = -b-a mixed II	a+(-b) = -b+a mixed I	-b+(-b) = -2b negative	b +(-b) = 0 Neutral	-c +(-b) = -c-b mixed V	*c+(-b)* *=c-b* mixed VI
c	-a+c = c-a mixed IV	a+c = c+a mixed III	*-b+c* *= c-b* mixed VI	b+c = c+b mixed V	-c+c = 0 Neutral	c +(c) =2c positive
-c	-a+(-c) = -c –a	a+(-c) = -c+a	-b+(-c) = -c-b	*b+(-c)* *= -c+b*	-c +(-c) = -2c	c +(-c) = 0

	mixed III	mixed IV	mixed V	mixed VI	negative	Neutral

In Table 5, the resultant directions of 12 vector time-fields from two 3-D bodies were calculated to find the result. Further, when interaction of two 3-D bodies occurs, there are at least three main kinds of areas taking place: (1) areas showing neutral directions of time values (and forces[1]) (0 = neutral), (2) areas showing positive directions of time values (and forces[2]) (2a, 2b, and 2c = positive1), (3) areas showing negative directions of time values (and forces[3]) (-2a, -2b, and -2c = negative 1). There are the other areas that occur as pairs as follow: (1) –b+a and b-a (mixed I); (2) b+a and –b-a (mixed II); (3) c+a and -c-a (mixed III); (4) –c+a and c-a (mixed IV); (5) c+b and –c-b (mixed V); (6) -c+b and c-b (mixed VI)

This mathematical result gave me the answer what my discovery in physics in 1987 was the formation of external magnetic poles.

Formation of external magnetic poles

Before I found the characteristic of the time dimension presented in this book, my discovery in physics had found

---

[1] Areas showing neutral-resultant directions of 12 resultant directions from 12 time-fields (or force-fields or poles) of two 3-D bodies.
[2] Areas showing positive-resultant directions of 12 resultant directions from 12 time-fields (or force-fields or poles) of two 3-D bodies.
[3] Areas showing negative-resultant directions of 12 resultant directions from 12 time-fields (or force-fields or poles) of two 3-D bodies.

accidently (Boonsri, 2012). The interaction of two bodies made me find the forming of external magnetic poles.

When the forming of external magnetic poles occurs, three main kinds of areas (poles) are always formed. **First**, the "South Poles", the pointing-inward directions of the North Pole of a compass, can be seen. The many heads of arrows pointing to them are shown in Figures 71 and 72. **Second**, the "North Poles", the pointing-inward directions of the South Pole of a compass, can be seen. Many tails of these arrows, pointing to them, are in Figures 71 and 72. **Finally**, the Neutral Poles, neither the pointing-inward nor the pointing-outward directions of the North and the South Poles of a compass, can be seen. See no heads and tails of arrows pointing to them in Figures 71 and 72. The reasons and mathematical results supporting my conclusion are at following details.

The external magnetic poles that were formed in the areas are relative poles. If we call the pole, which the south-end of a compass needle is pointing to, the "North pole or Positive pole or Like pole or pole A or the component of the pole B or the relative pole A", and then we must call the pole, which the north-end of a compass needle is pointing to, the "South pole or Negative pole or Unlike pole or pole B or the component of the pole A or the component of relative pole A" respectively. The last, we have to call the area, which is not both of the north-end and the south-end of a compass needle pointing to, "the neutral area"; this generates the force which repulses two ends, the north-end and the south-end of the compass needle. (It is the area of zero force.) The interaction of two

3-D bodies will always form the resultant directions of 12 time-fields.

In the area of my discovery in physics in 1987, the interaction between two time-fields occurred. The first one is the pyramidal shape make of paper attached with electrolytic capacitors at the steep slope that my friend and I made. Another one is a time-field inside the earth. See Figure 74. In addition, the earth is rotating around the sun, so two time-fields are rotating around the sun as well. Present scientists and engineers know very well the characteristic of many kinds of magnet and how to use its characteristic to invent a dynamo to produce electricity. However, they have never seen what happen before the poles of magnet take place. The formation of external magnetic poles indicated that the formation of magnetic poles will occur after interaction of 12 time-fields (or 12 force-fields) occurs. There are three kinds of areas as the main results: (1) relative North Poles, (2) relative South Poles, and (3) neutral areas.

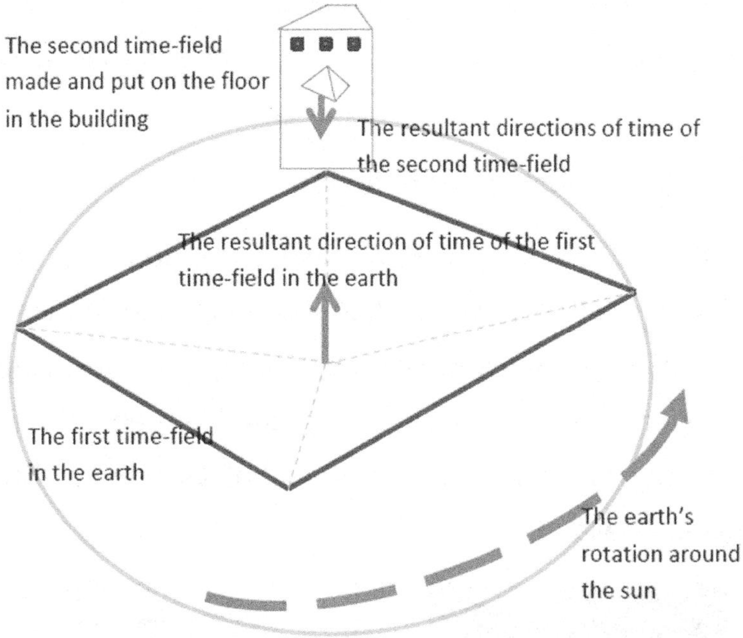

Figure 74: The interaction between two 3-D bodies that are on the earth rotating around the sun. The first one is the time-field in the earth and the second one is the time-field made by my friend and me.

In the figure above shows the interaction of two 3-D bodies, which were the main cause of my discovery in 1987. The first 3-D body is the time-field, which I made in previous report. The second 3-D body is the time-field of the earth. To prove the formation of magnetic poles, it is possible to carry out an experiment even in your house. Notably, the floor that was used to do the experiments was made of wood mainly from trees and cement floor of the constructed building in my university. The results of the experiment are available on YouTube <https://www.youtube.com/watch?v=kNPxjpef9Ys>.

Conclusively, the strange phenomenon in physics in 1987 which I discovered was the formation of external magnetic poles.

Duplication to test the conclusion

In 2001, the first of two duplicating was done by my two students in that time, Mr. Prakob Kongnual and Mr. Pisatesit Sirichokchatchawan (See Figures 75 and 76 respectively), on the wooden floor of the rent room of the Horruen dormitory in front of Bansomdejchaopraya Rajabhat University (BSRU), Bangkok, Thailand.

Figure 75: Prakob Kongnual, an alumni of BSRU	**Figure 76: Pisatesit Sirichokchatchawan, an alumni of BSRU**

The second by both students and some students of the lecturer of physics, Pongsak Narksuwan, Faculty of science and technology, on the cement floor of the room number 779 now of the new constructed building called “Building No.7” in BSRU. The same results as the initial discovery occurred. At this moment (January 2017), the magnetic lines on the floor have been appearing since 2001.It is possible to notice that the forming of the external poles was created on both the wooden floor and on the cement floor of a new constructed building. Scientists may argue that on the cement floor, there are many steel bars which are tied in many shapes for construction, but no one has any argument in the case of the wooden floor because of the characteristics of wood which cannot be attracted by a magnet.

One of the implications of the experiment is that the formation of the relative poles from a wooden floor (from trees) (See Figure 77) is faster than that conducted on a concrete floor (See Figure 78). The formation of the north (N), south poles, and neutral areas from a wooden floor took 2-3 hours; however, it took 2-3 days when the experiment was conducted on a concrete floor. In addition, before the experiments were done, the magnetic lines on the floor were scanned but the results did not show any formation of the relative poles. Besides, the compass needle pointing to the North geomagnetic pole of the earth was normal. Overall, the results of these experiments produced the same results about the formation of the north, the south, and neutral poles of magnetic field in the room I discovered in 1987. More information can be obtained from
https://www.youtube.com/watch?v=kNPxjpef9Ys.

The magnetic lines on the wooden floor of the rented room of Mr. Prakob Kongnual in the Horruen dormitory in front of Bansomdejchaopraya Rajabhat University, Bangkok, Thailand in 2001.

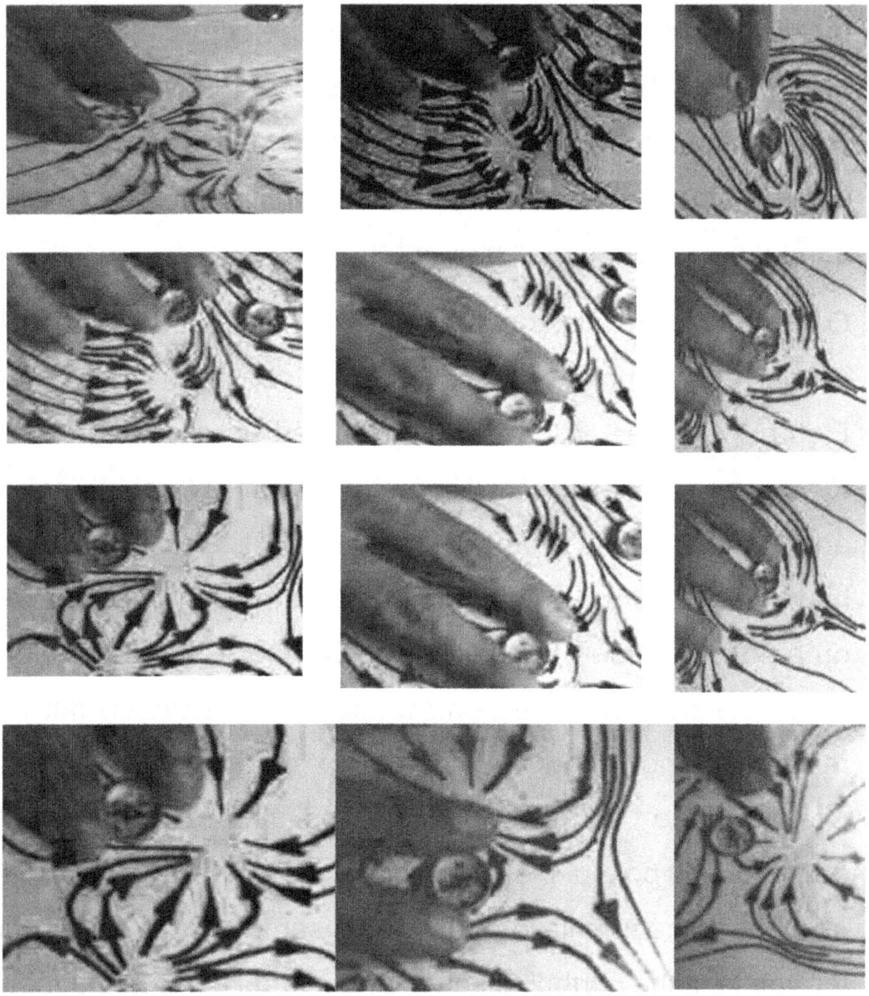

Figure 77: The formation of poles on the wooden floor of Prakob's room in 2001.

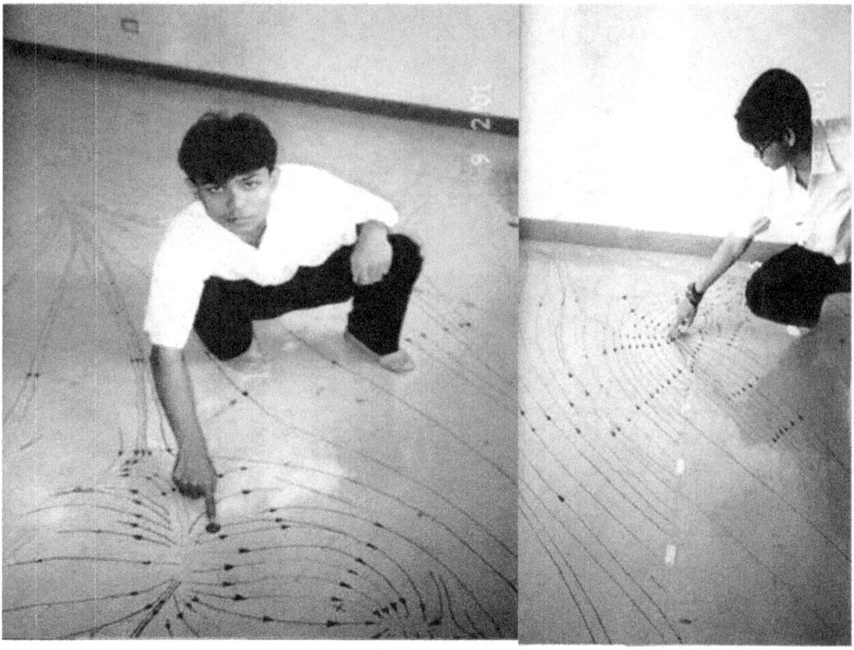

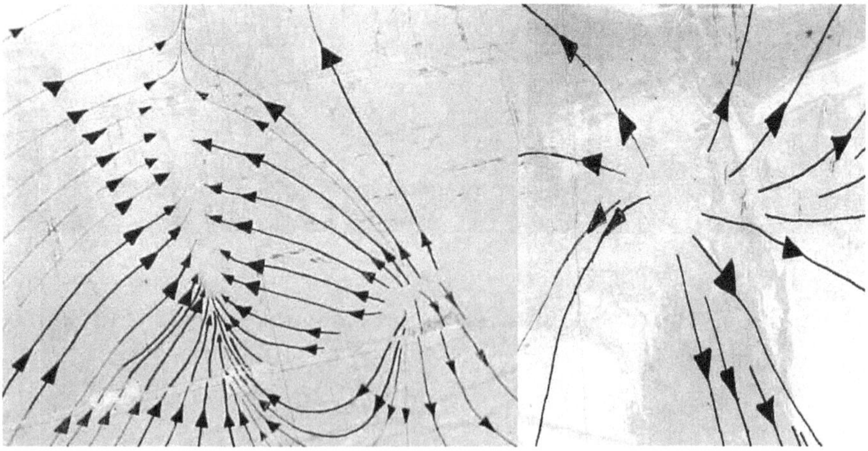

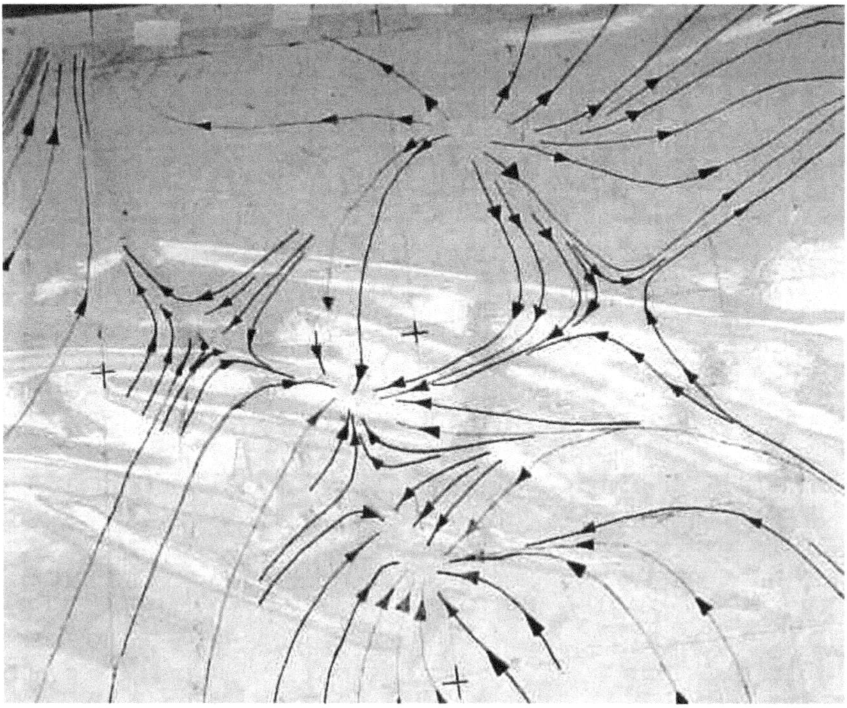

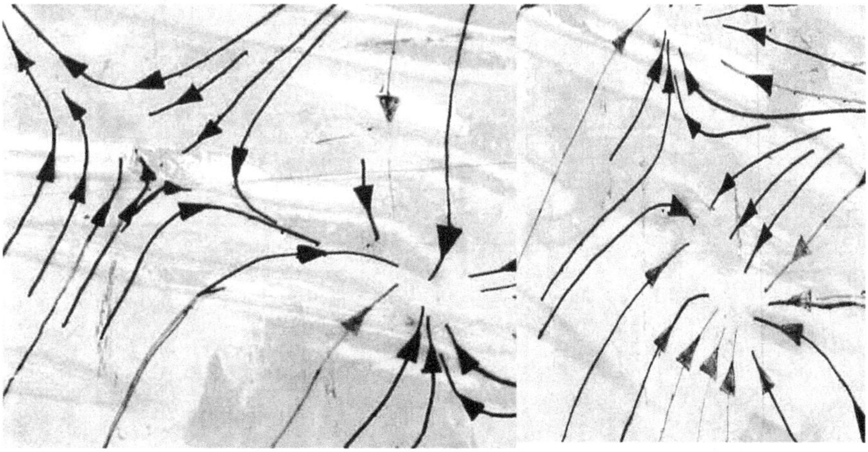

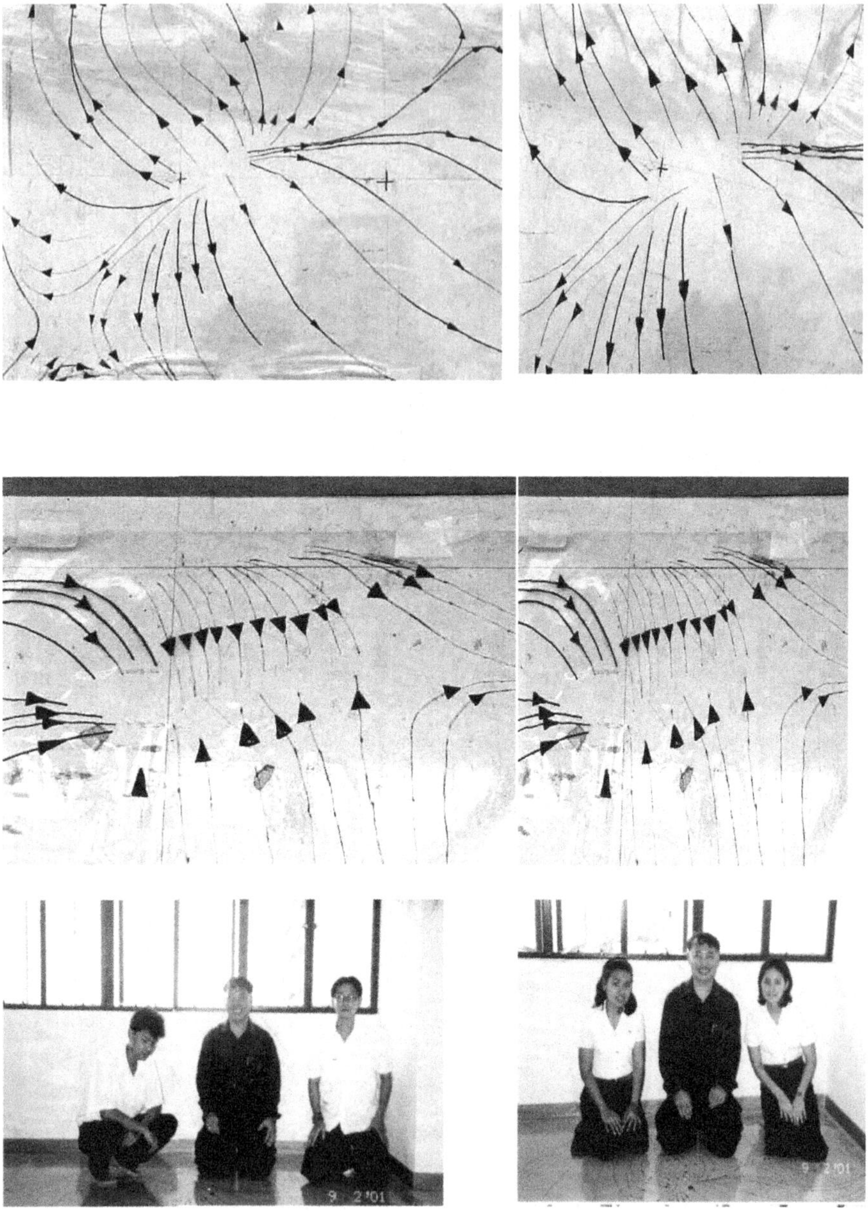

Figure 78: The formation of external magnetic poles on cement floor of new constructed building that no one used it in 2001.

Application of interaction of two bodies: unsolved mysteries in physics solved

From the experiment that was conducted, we found that the time dimension is at subatomic level. Having considered time in the context of subatomic level that has many particles, we began to imagine the formation of poles in subatomic level. In other words, when two particles are rotating and they are near each other, they are bound to form at least three main kinds of areas in subatomic level.

The formation of external magnetic poles can help scientists to understand the reason why the enormous magnetic field occurs when an atomic bomb explodes, and how and where the magnetic field comes from. In subatomic level, there are uncountable amount of particles that their interaction take place. First, particles always form at least three areas of forces; positive, negative, and neutral. When the formation of magnetic forces (or electromagnetic forces) was destroyed, the enormous magnetic forces from particles were released.

Moreover, the formation of external magnetic poles helps astrologers to imagine three different areas when two galaxies (two 3-D bodies) rotate or move near each other. Specifically, a black hole is one of three areas that show neutral area of force.

The 36 resultant directions of time of two 3-D bodies in subatomic level can show the characteristic of waves while two bodies are spinning. One of the greatest unsolved mysteries in physics is about rays of light could be a wave or a particle ("10 Greatest Unsolved Mysteries in Physics", 2016). Sometimes rays of light create interference patterns that are typical of waves by reflecting off surfaces, but sometimes, they show the characteristics of particles that are being used to liberate electrons from their shells. Physicists could not answer how light determines whether it acts as a particle or a wave or both at the same time.

One of the characteristics of wave we know is that; when two sources making waves interact, there are three main areas (interference pattern) which forms a resultant wave of greater, lower, or the same amplitude. The characteristics of a particle include its length, width, and thickness. If a three-dimensional body is a photon, interaction between two photons spinning can solve one of the greatest mysteries in physics is about rays of light. In other words, light acts as a particle when it is considered as one of a three-dimensional body that consists of length, width, and thickness. In addition, light acts a wave when there is an interaction between two photons that are spinning with both the particles and waves are considered at the same time. This finding may be one of the linkages between quantum theory and Einstein's theory that are working together.

The polarity of gravitational force

Because the time is in all dimensions including in force, so there are forces at eight resultant directions of time of the eight parts and the forces in the six resultant directions of the six time-fields of the three-dimensional body as well. I named the forces running from the center to outside of the three-dimensional body "Send-Out Force." See Figure 79.

When time is a vector; as a result, the direction of time dimension is important. With the law of directions of time, different directions of time can show different kinds of poles.

For example, if a three-dimensional body is a magnetic bar, the north and the south poles in the magnetic bar are performing the attracting force without any repulsive force, but the forces of the two poles are different from each other. In other words, the North Pole is different from the South Pole.

When a north pole faces a south pole, we feel a force pulling the magnets together. But when the North Pole faces each other or the South Pole faces each other, we would have felt a force pushing them apart. The opposite poles attract (pull together) while similar poles repel (push apart).

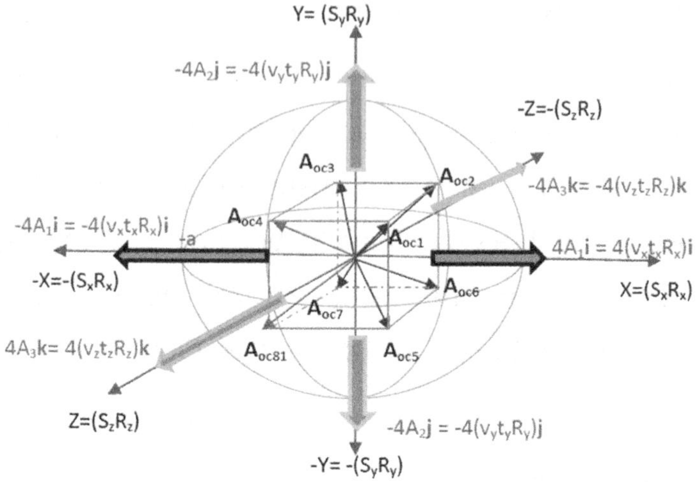

Figure 79: The time dimension in everywhere including force. Six resultant directions of six time-fields (or six force-fields or six poles) running from the center to outside of the three-dimensional body such as a magnet bar

The figure shows the resultant directions of six time-fields running from the center of 3-D body. The resultant directions are different from each other. Notably, when time is in force, then the directions of time will show the directions of force too. However, in the event gravity has opposite directions with all resultant directions of six time-fields, the gravity at different sides of the earth will be unique. Thus, it was concluded that the gravity has its own poles (or there is the polarity of gravity).

The invisible area around a magnet which attracts another object is called a magnetic field. Magnetic objects such as paperclips will be pulled towards the magnet if they are placed in this field. There is the time dimension in a magnetic field as well, so if I call the upper time-field of a magnet "an upper magnetic force-field," then the bottom time-field of a magnet has to be called “a bottom magnetic force-field.”

If we call the upper time-field of the bar called the North Pole or the positive pole or the like pole, then we have to call the bottom time-field of the bar called the South Pole or the negative pole or the unlike pole respectively because the two names have to be called in the opposite direction.

With time as a vector, there are six time-fields in minimum in a three-dimensional body, so there are six force-fields in minimum in a three-dimensional body as well. These mean the magnet bar having six poles not two poles in minimum as well.

Moreover, each field has also its own characteristic. The upper time-field is different from the bottom one. The magnetic force at the upper and the bottom sides of the magnet bar is stronger than that at the four rest sides. If the North and the South Poles are the first group that is called “Dominant”, then the rest will be called “Recessive.” Not only the force at the North and the South Poles of the bar magnet is stronger than the other four poles, but all sides of the bar magnet have their own characteristics.

The different resultant directions of the time-fields show the different resultant directions of the force-field (and poles). In other words, the time-fields show the force-fields. The force-fields show the poles. Each of the time-field shows the monopole and one characteristic. All of them are the vector fields and are relative not the absolute values.

We could not see the different characteristics between the North and the South Poles, if only one magnetic bar is held. We may easily misunderstand and conclude that the forces at the two ends of the bar are the same attracting force because both poles can attract iron filings. Making us know the difference between the two poles of the bar is the second magnetic bar. The repulsive force inside the magnet bar will be shown when the North Poles or the South Poles of two bars faces each other.

In the minimum case, the six resultant directions of the time-fields (or force-fields or poles) are sending out from the center of the three-dimensional body.

Present physicists believed and conclude that the gravity has the only attractive force. If the three-dimensional body is the earth, and the gravitational force at the upper force-field with its own characteristic is different from that at the bottom force-fields, then the gravitational forces in the different sides of the earth are not the same characteristics too. In other words, the gravitational forces at the North and the South Poles of the earth are not the same. They have their own characteristics. As a result, **I concluded and predicted that the gravitational force has its own poles.**

The repulsive force of the gravitational force

Moreover, when the gravitational force at each pole has its own characteristic, then the repulsive force of gravity can be forced to appear to anyone especially scientists. For example, if they want to see the specific characteristic of the gravitational force under their feet, they have to invent something that can produce the same characteristic of the gravity on the ground.

In the past, a Russian inventor, Viktor S. Grebennikov, brought some parts of insects to produce the same poles between the floor and his self-lifting vehicle, and his vehicle can move up from the ground without the work of any machine (Keelynet.com, 2016).

In the future, I predicted that the revolution of constructing in this world will be taken place. We will see floating buildings in the air around the world and those buildings will change all of life styles of mankind forever. Right now, the cost of the land in the city is higher than that of in the rural area. We have to spend very high price for very high constructed buildings, in the future, the cost of the construction will be cheaper. In other words, the unlimited change of the world of construction is beginning at this point.

Forces from different poles of gravity as new-future-cheap power

At present, we use petroleum products that will run out at the middle of this century. We use in many ways to produce energy and forces. With the polarity of gravity, flying cars without wheels will be invented for people to travel to the area we want to go including the last frontier of humankind in the earth, the ocean.

When the gravitational force has its own poles, we can produce electricity for electrical devices. For example, if t = 1, the direction of the gravitational force running from the center to outside of the earth in a "new engine" and at t = 2, the opposite directions of the force running from the outside to the center of the earth were controlled in the engine, we can use it to produce electricity and solve crisis of energy that will take place before petroleum products run out.

Right now, the rockets using the chemical reaction from Hydrogen and Oxygen are being used for astronauts to orbit around the earth. The application of the polarity of gravity can generate new energy for humankind to launch the flying objects. While a space ship is making the same kinds of force of the ground, it will be lifted from the ground. However, at that moment, the moon is attracting the earth with different poles of the earth as well. As a result, the ship will jump up from the earth to the outer space with enormous speed because the ship is repulsed by the gravity force of the earth and is attracted by the gravity of the moon. The new frontier in space will have been explored after the new energy is used for future transportation.

The new energy can solve our many problems directly and indirectly. For example, the huge lifting vehicles with the size of a football stadium bearing million gallons of water without any problems about weights will be invented to deliver water from lakes to very far areas in deserts. The large deserts which no plants can grow now will be cultivated. The water in places that people do not want will be delivered to the other suitable areas for cultivation as well. The new technology will make the plants grow faster for solving starvation of poor people around the world. The tourism in space to the moon or to the farthest planet in the solar system will cost very little because space-ships launch to the sky very easily and rapidly beyond imagination of present people. In addition, we and aliens visiting the earth will have the first formal contact.

In the air of the earth, many floating buildings will be constructed to move away from disasters such as flood, earthquake, hurricane, and Tsunami. The revolution about forces helps us to invent cheap-portable generators for using everywhere in the world. The expensive nuclear power plants with nuclear waste and pollution around the world will be rejected.

The problems of global warming will be solved because the space-ships will be used to fix the hole of earth's atmosphere smaller. The storms such as Hurricane or Typhoon decrease their velocities because the enormous space-ships create an opposite direction of force to slow down the speed of natural storms. The space stations between the earth and other planets will be constructed easily with low cost and be launched to the accurate positions on a space map. The exploring of the planets in different galaxies will begin. Humankind's technology of travelling to other planets will be developed for people around the world to see the houses of many present aliens who are sightseeing the earth.

The people in the United State of America will visit the countries at the opposite side of the earth such as the Southeast Asia or Australia in a few minutes because the vehicles jump out to the outside space of the earth and land easily to the other side.

By product of the poles of gravitational force

In a three-dimensional body, there are six time-fields in minimum. A time-field shows one force-field which will show a pole. The time-fields in the three-dimensional body indicated that all three-dimensional body including a magnet have at least six time-fields, it is the reason why we could not find a magnet bar with the monopole in nature. Right now, physicists find a one-pole magnet only in the quantum mechanics theory, if they want to see it in nature, they have to make the time-field which is the pyramidal shape by using material such as hard paper and wood because the pyramidal shape is the "monopole magnet."

Moving without angular momentum

If a three-dimensional body is a particle, then the six resultant directions of force of six force-fields in all particles are perpendicular from each other; these mean that we can control the force in one of the six perpendicular directions in atoms and send it out from the particles to outside space. The vehicle with this controlling will move without angular momentum. While it is moving to the south, suddenly, it can change the direction from the south to the west.

Gravity running from the center to outer space

In this topic, I want to show the different views of the direction of gravity between physicists and me. While scientists are teaching the directions of gravity running from outside space to the center of the earth, I am telling that gravitational force is pointing from the center to outside of three-dimensional body. In the future, scientists may do some experiments to answer this topic; only two possible ways of answers will get: (1) present scientists are still right, (2) my answer is right. However, the polarity of gravity still is the same. In other words, the direction of gravitational force will not affect the poles of gravity (Boonsri, 2016).

When we see the magnetic bar attracting the iron filings, we see the iron filings moving only, but we do not see the attracting force, which is grabbing the iron filings. While the iron filings are running to the pole of the magnet, the force in the magnet is sending from the center of the magnet to outside space. The direction of the iron filings moving to the pole is the opposite direction of the magnetic force. It looks like the iron filings "falling" to the pole of the magnet. Acting in the opposite way of the iron filings running is the real direction of the magnetic force.

Time is in force, so the directions of time can show the direction of force as well. We can conclude the directions of values of time presented in the second chapter to fifth chapter in the Table 6.

Table 6: Time directions presented in the second chapter to fourth one

No	The items	Their directions
1	The values of time matching with numbers in a number line	are running from the center.
2	The values of time in the N lines in maximum in the three-dimensional body	are running from the center.
3	Directions of time along the N number lines or curves	are running from the center.

4	The values of time in three principal axes	are running from the center.
5	The values of time in the six directions along the three principal axes in the three-dimensional body of these axes	are running from the center.
6	The values of time in eight parts of the three-dimensional body	are running from the center.
7	The values of time in all points in all eight parts	are running from the center.
8	The eight resultant directions of time in these parts	are running from the center.
9	The six resultant directions of time of six time-fields (or six force-fields or six poles)	are running from the center.

In Table 6, having considered from (1) to (9) that shows the resultant directions of the six resultant time-fields, the six force-fields, and the six poles, we found in the three-dimensional body, these resultant directions are running from the center of the body.

If the three-dimensional body is the bar magnet, then the resultant directions of the magnetic forces at all poles are running from the center of the bar to the outside space.

If the three-dimensional body is the earth, then the resultant directions of the magnetic forces at all poles are running from the center of the earth to the outside space.

There are values of time in gravitational force too. With F = mg of Newton's equation, we can find the time dimension in the symbol g, which represents the acceleration. With these previous reasons, **I concluded that the gravitational force is running from the center of the earth to outer space.**

A magnet is one of the three-dimensional bodies. Iron fillings are running to the North and the South Poles in the opposite directions of the force in the bar. Similarly, when something was falling to the ground, we did not see the gravitational force but we saw the objects falling. The direction of gravitational force is acting in the opposite direction of the things falling. Its characteristic is as same as the characteristic of a magnet bar.

In other words, gravity force has been sending its force from the center of the earth to grab everything to the center. It is trying to pull the things down to the center of the earth. When Newton saw the apple falling from the tree, he did not see the force, but he saw the apple falling after the apple had reacted to the gravity force of the earth.

We can notice that the "Send-Out Force" presented at the first part of this chapter is the gravity. Time in both gravity and subatomic level are the different views that may unite the tiny world in subatomic level and the big world in the sky together.

The directions of magnetic forces of magnet and magnetic field of the earth

Scientists believed that the direction of magnetic force of a bar magnet and magnetic field of the earth running from the North to the South (dominant poles). However, when the formation of the external magnetic poles was discovered, it indicated that not only the direction of magnetic force running from the North to the South but also the direction of magnetic force running from the South to the North.

The discovery in 1987 and two duplication in 2001 showed that the directions of magnetic force in a three-dimensional body running from the areas as follow: (1) the North to the South or the neutral areas (many tails of arrows linking to many heads and neutral areas), (2) the South to the North or the neutral areas (many heads of arrows linking to many tails and neutral areas), and (3) the neutral areas linking the North and the South areas together (neutral areas linking to many tails and many heads of arrows).

Summary of the chapter

Present scientists believe that the gravitational force is different from the other three fundamental forces of nature because it has only attractive not repulsive. Physicists could not unite these four fundamental forces into one coherent picture because the gravitational force could not link to the others.

The polarity of the gravitational force is challenging scientists to prove it in the future. The force has its own poles that were proved mathematically before this book was written. The view about the direction of the gravitational force between physicists and me is different from each other. Present physicists explain the direction of the gravitational force running from the outer space to the center of the earth, but I found many reasons indicating that the direction of the force is running from the center of the earth to outside space and trying to grab things down to the center of the earth. While we were looking at something falling from the sky, we saw the things and the falling direction; we did not see the gravitational force. In other words, we saw things, but we could not see the gravitational force. It looks like looking at the iron filling running to a pole of a magnet because we saw the iron fillings; we could not see the magnetic force.

The poles of the gravitational force will be used to be the new energy for us to travel to the other planets in the further galaxies in the future. We are witnessing the genesis of the future physics.

After having found many characteristics of time, I found that pyramidal shape is the shape of the time-field in three-dimensional body. The secret of power of the pyramidal shape that was written in the past was opened up in this book. The pyramidal shape has its own characteristic; its characteristic looked like "strange" until we knew or understood it. Similarly, before twentieth century began, a black body was so strange for scientists. After having found the quantum theory, we did not look it as "a strange thing" anymore.

I discovered the formation of external magnetic poles in 1987 and answered what happens before a magnet bar forms its poles. Scientists know a magnet attracting iron fillings; however, they did not know what happen inside a steel bar. The two dominant force-fields have different characteristics from each other because of different directions of time. My discovery indicated that the formation of external magnetic poles has at least three main areas: positive areas, negative areas, and neutral areas.

Einstein's first Relativity Theory written in 1905 was accepted after a well-known scientist at that time had presented his work that was similar to Einstein's result in 1909. Planck's equation was accepted many years after conservative physicists had changed their views in the black body. I really did not know when my research will be accepted in the future. However, this book is scientific evidence of existence of time dimension in subatomic level, a pyramidal shape as a time-field with monopole, the formation of external magnetic poles that always forms at least three main areas, and the new characteristics of gravity. All of them were proved mathematically.

Mathematics is one of scientific method to solve the truth in nature. Einstein use mathematics to study nature. Many secrets in nature especially in physics were found by him with his equations. No one believed his results at the first time; however, Einstein challenged scientists to prove his results about nature. This book is in the same boat of Einstein's work because applied mathematics is used to prove all results concerning the time dimension in this book.

When I started to study about time, I found the situation called "paradigm shift of knowledge of time." In other words, present knowledge about time did not give me any beneficial results to answer my discovery. Without books, persons, experiments (except Minkowski's 4-D model) to guide how to "catch" the time dimension to study, I stuck this pain point for nearly 12 years before I found the first answer in 1999. Finally I found that mathematicians use numbers to represent everything, and time is in everything, so I began to prove mathematically the existence of time in numbers that was one of the most difficult steps. If this book will not be accepted to publish, and someone, a target audience, wants to know about time, he or she will face with one of the most mysteries in science (I have found) and will lose his or her time like I did to search for what and where time is.

No one knows time in numbers. Present scientists including Hawking believed that time is the fourth dimension separating from numbers as (x, y, z, t) following Minkowski's 4-D model. We always overlook time that we spend when we draw a line. In other words, the values of time are always in the line. My mathematical proof by using the Newton's concept of moving in the first step shows time matching with all numbers in three principal axes (x_t, y_t, z_t).

And I really hope that my research will be benefit for some scientists to understand the whole of nature deeply and widely in the future, and will be used to solve energy crisis before petroleum products run out. The very high prices of energy may be suitable for some small group of people in all countries if scientists cannot solve the energy problem before energy crisis begins. The big countries that have the enormous size of population such as China and India will face to big economy and social problems most. When income of countries is used to buy the high price of energy, many social problems will increase rapidly following economy problem. My research may be one of ways to solve the energy problem in the middle of the century.

The formation of poles on wooden floor pointed out that knowledge about insulation may be improved. Moreover, three main areas that are formed when the formation of external magnetic poles occur warned us that the theory of electrons creating magnetic field may be developed because it is supposed to cover my discovery in 1987. If the time equation explaining many equations in physics is a part of the whole of Theory of Everything, it is not the end of studying physics but it may have just been the beginning of new physics for our children in the future.

This book was written by using applied mathematics, so it concerns directly STEM. Time is a basic knowledge in the worlds of science and social science. The book can be used for courses as follow:

In a short-run, this book may be used for advanced mathematics, advanced physics, advanced chemistry, and advanced engineering and technology.

In long-run, when people know well about time, this book is studied in general science as follow: general science, general engineering and technology, general mathematics, general physics, general chemistry, general logistics and supply chain, and general economics (the new knowledge about time that can link micro and macroeconomics together that economists including Alfred Marshall and John Maynard Keynes have never done before).

Chapter 7

Physconomics: The Physical and Economic Properties of Time

We build too many walls and not enough bridges.

Isaac Newton

The poverty in the theoretical economics was written by many economists in many countries for a long time. For example, Robert Kuttner written his article "The Poverty of Economics" where he indicated that in present economic theories, there were some weak points that economists were supposed to think about (Kuttner, 1985). In addition, in his article, Kuttner referred to also Robert Heilbroner, the economic historian, that new theories in economics would explained questions as follow: what a price is, what money is, what killed full employment, and what the boundary between economics and politics, or society is (Heilbroner & Milberg, 1995). In other words, the questions may help economists to see their directions of development for their real world theories with more accuracy in the future.

Thorstein Veblen, one of the well-known American economists, gave his comment on the development of the economic theory. There had been a million words of some economists which were written to point out the weak points in economics, but if there is no one devoting his (or her) time for searching the new theory or new alternative in economics, then the beginning of the development in economics could not be taken place (Veblen, 1898).

Karl Mark, well-known economists, analyzed that if the two forces which had the opposite directions of the theoretical demand and the supply curves acted at a point called “the equilibrium point”, then all transactions in economy would stop. John Stuart Mill rejected to use the model of the theoretical demand and the supply curves to explain the phenomena about prices because he viewed that without considering about the culture of the society, the theoretical price was explained by the demand and the supply being inappropriate. (John Stuart Mill quoted in Guerrien 2002).

In present, one of the new groups is called "real world economists" are trying to develop economics with new ideas that can explain the real world economics because they saw the present neoclassical economic theory that could not use to explain the real world economy. For example, in microeconomics, assumptions in the theories are statics and no any parts of the theories show the difference between statics and dynamics. (Paecon.net, 2004). Besides, in terms of macroeconomics, Keynes' theory could not explain unemployment and inflation in 1970s, which occurred simultaneously (Willes, 1981).

In the past, Alfred Marshall used the deductive method to create the neoclassical microeconomic price theory. In present, mainstream economics still explain economy phenomena with normative economics, not positive economics. In addition, neoclassical microeconomics created by Alfred Marshall and macroeconomics created by John Maynard Keynes explaining what happens in economy phenomena in different levels; however, both of them still used the concept of demand and supply curves as fundamental concepts (Boonsri, 2014).

Toward the next economics

In 1981, Peter Drucker wrote his article "Toward the Next Economics" and predicted that next economics theories should be able to link these microeconomics and macroeconomics together because present two levels of study in economics could not link together ("Peter Drucker - New World Encyclopedia", 2016.; Drucker, 1987). Until now, no economists including Alfred Marshall and Keynes have linked these two levels together.

Time is still crucial guideline for me to search for a new theory created with positive economics for explaining the real world economy. With the different paradigms of Marshall, the inductive method was used to create the new model of the price in the real world. The price system in the real world consists of buyers and sellers, and the buyers have money to buy goods and services of the sellers. The sellers have goods or services to sell for the buyers, the sellers want to receive money from their selling. With Locke's comment, a price came from the concept of the demand and supply.

At t = 1, the time before exchanging, the buyers have money and the sellers have goods or services. The sellers make the price of their goods and services, in other words, the sellers are price makers, and the buyers are price takers. The price is data, which the sellers are telling the buyers how much money buyers have to pay for the sellers' goods and services. A buyer's feeling on a price may be different from other buyers.

Some of them feel that the price is too high; some feel that the price is high; some feel that it is a reasonable price; some feel that it is cheap, and some feel that it is too cheap. If the price of the sellers is too high for all buyers, no buyers pay for sellers' goods. In the view of the most sellers, they love to sell their goods with high price. In contrast, in the view of the most buyers, the buyers love to buy the seller's goods with low price. The feeling of the sellers and the buyers had not been discussed in this book.

At t = 2, the moment of exchanging, the price as data, which the sellers are telling the buyers, is $ 10 per each of a shirt, and if the buyers want two shirts of the sellers, the buyers have to prepare their 20 dollars for buying; this is the specific form of the price. The conclusion is that the general form of new mathematical price model is as the following topic.

New price model with time

The new mathematical price model in microeconomics and macroeconomics that I called as the new price function with time that can be written as follow (Boonsri, 2014):

$$P_t = f\,(M_{Tt}, Q_{Tt}) = \frac{M_{Tt}}{Q_{Tt}}$$

The above equation is considered to be the most important equation of new economics. It shows that the price in the real world is data. That is, in a price system when sellers sell their products to buyers, they will tell the buyers how much money they are supposed to pay for the products. In the process, money will be transferred from buyer to seller; consequently, the products will be transferred from seller to buyer. At the moment of exchange, the price is the only data. Otherwise, the price does not come from quantity demanded and quantity supplied as articulated in Marshall's price theory in microeconomics of present mainstream theories.

The letter "P" in the equation represents a price of the product per unit; "M_T" is the "accumulation" of buyer's money, which is transferred from the buyers to the sellers; "Q_T" represents the "accumulation" of the quantities of the product, which is transferred from the sellers to the buyers; and the subscript "t" represents the values of time matching with all values of P, M_T, and Q_T. For example, if we have two dollars (M_T), for buying two cans (Q_T), then the price is 1$ per can, this activities have taken place at a time valued as t.

With this new price model, all four main factor costs can be explained by the same model. For example, "interest" is one of prices as data that owners of money tell borrowers borrowing money how much money of borrowers has to be paid for time of using owner's money. "Rent" is one of prices as data that owners of property tell renters how much money of renters has to be paid for time of using the owners' places. "Wage" is one of prices as data that employers tell employees how much money of the employees will pay for employees' time of working for employers. "Profit" is one of the prices as data that entrepreneurs tell themselves how much money of them has to pay for entrepreneurs' time of their working.

This price model can be used for explaining four real world economy activities: consumption, production, exchanging, and distribution. For example, the producer E pays average 4 $ per unit for a piece of cloth for making a shirt, 1 $ per unit for the wage of the worker B, 2 $ per unit for the place of the landlord C (rent), and 3$ per unit for the interest of the banker D. The total cost of the producer E is 10 $ each, (4+1+2+3). The producer E makes the first price as 20 $ per unit for selling of the goods. He or she gets profit 10 $ each in his or her pocket.

The cost of the wholesaler F is 20 $ each, if the wholesaler wants profit as 5 $ each and his or her costs 5 $ each (for wage, rent, interest), so if he or she makes a new price at the second step, $ 30 each, and sell the goods to the retailer G.

The cost of retailer G is 30$ each, the retailer G wants profit as 20 $ each and his or her costs 5 $ each (for wage, rent, interest), so if he or she makes a new price at the third step, $ 55 each, and sell the goods to the end user or buyer H.

This example shows that the price as 55 dollars has many dimensions hiding in it. There are also four economy activities in this example: the producer E acts production, exchanging, distribution. The wholesaler F and the retailer G, both of them act exchanging and distribution. The buyer H acts exchanging and consumption.

A price in the real world is so complicated because the time equation points out that a point can have N dimensions in it in maximum.

In addition, the new price model is made by using of the demand side as the amount of money, which is transferred from buyers to sellers, and the supply side as quantities of the product, which are transferred from sellers to buyers. This concept came from positive economics, studying for an answer "what is it?" and inductive method that I observe real world prices for creating a new theory. The concept can be explained in three-dimensional figures.

The axis of the M_{Tt} shows the accumulation of buyer's money. If the axis is longer, then the money increases. The axis of the Q_{Tt} shows the accumulation of the quantities of the product. When the axis is longer, then the quantities of the product increase as well. See Figure 80.

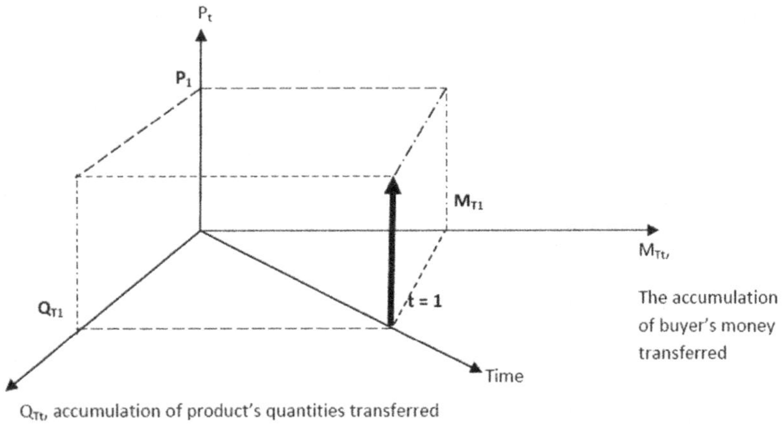

Figure 80: Three-dimensional figure explaining how to show a price in the real world.

The figure represents a price in the real world. The data shown in the figure is about sellers telling buyers how much money they have to pay for the products. Besides, price is shown to have several hidden dimensions.

The price has many dimensions hiding in it such as all costs and profits of many merchandisers. See Figure 81. For example, the price of a bottle of drinking water, P_1 as two Euros consists of many dimensions such as fixed cost, variable cost, and tax. **Fixed cost** per bottle consists of the rents that the producer A, the wholesaler B, and the retailers

C pay. **Variable cost** per bottle consists of the price of the plastic bottle, the plastic cover, and a sticker on the bottle, the water in the bottle, the wages, the interest, the profit and the tax.

With the new model, the cost of firms consists of many dimensions at different values of time. In the real world economy, many kinds of cost such as data transferring, advertising, setting events for marketing and official training for newcomers of different organization are occurring in supply chains of a product and are making significantly higher costs of sellers. In addition, sellers will increase prices of their products if their costs are higher. Nobody knows and receives the information of details of markets completely. In other words, four main factors in present theoretical economics: the rent, the wage, the interest, and profit were developed for real world economy system because of the existing of the many kinds of cost such as natural resources as raw materials, chemical substances, the process of production which may have hundred steps with technology, and transportation.

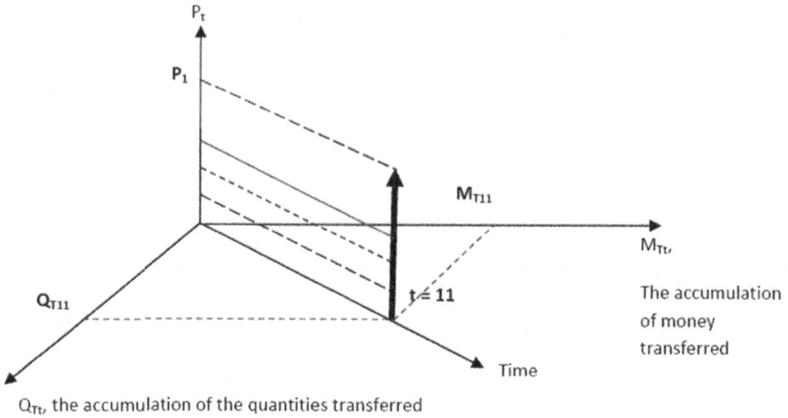

Figure 81: The price P1 consisting of many dimensions and the time links all of the markets together.

The price in the real world is shown in the picture to many hidden dimensions. The hidden dimensions may show different markets from various aspects of a similar product. In the model, each real world market is represented by a unique characteristic. The market of shirts, for instance, is different from the market of houses, the market of energy, and the market of books. In addition, the shirt market of the country A is also different from that in the country B such as different rate of tax.

Buyers in each market have different tastes, different life styles of people, different culture, different religions and etc. The same advertising, for instance, affects people in the different ways: some buys, some wants to buy, some refuses to buy. Most of the buyers are price takers, and most of the sellers are price makers. Each family has the different ways of making decision about buying. The model shows a link between microeconomics and macroeconomics by using the accumulation of the whole natural resources, P_1, the whole fixed cost especially the rent, P_2, and the whole variable costs, P_3, such as the whole wages, the whole interest, and the whole profit together. See Figure 82.

Exchanging line: new paradigm of demand and supply

With new paradigm of the time dimension, we can see prices hiding many dimensions in the different values of time and create a line which I call "an Exchanging Line" to explain prices in the real world. See Figure 83.

For example, at t =1, average price of a bottle without its cover is P_a that the producer A bought it from a bottle factory. At t = 2, new average price P_b that came from $P_a+P_{cover} = P_b$, that the producer A bought some covers to use with empty bottles. Each of a cover is P_{cover}. At t = 3, new average price Pc that came from $P_b+P_{sticker} = P_c$ that the producer A bought some stickers from a sticker company. Each of a sticker on a bottle is Psticker. At t =4, new average price P_d that comes from $P_c + P_{water} = P_d$ that the producer A bought the water from the company delivering mineral water. Each of water in a bottle is Pwater. At t =5, new average price P_e that comes from $P_d + P_{wage} = P_e$ that the producer A paid for his workers per bottle. Each of wages for a bottle is P_{wage}.

In macroeconomics, M_T can represent all money to buy all bottles of drinking water of the whole market in a country or countries, Q_T can represent the number of all bottles of drinking water, and then P_1 can represent the average price of a bottle of drinking water of the whole market in a country or countries at any time. Exchanging Line with time is also the linkage between microeconomics and macroeconomics as well.

The example shows that the time dimension links all of the markets in the different levels of economy.

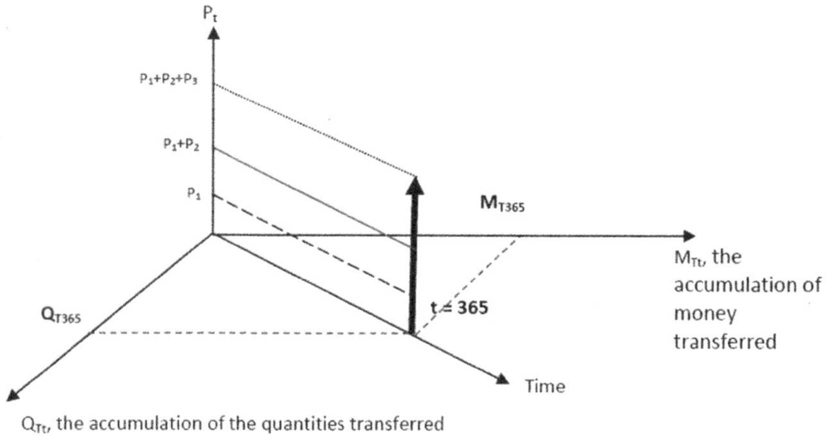

Figure 82: The time linking microeconomics and macroeconomics together with all of the markets in an economy system

The figure shows how time can link micro-and macroeconomics together. This is the specification of the next theory of economics following the prediction of Peter Drucker in his book "Toward the Next Economics."

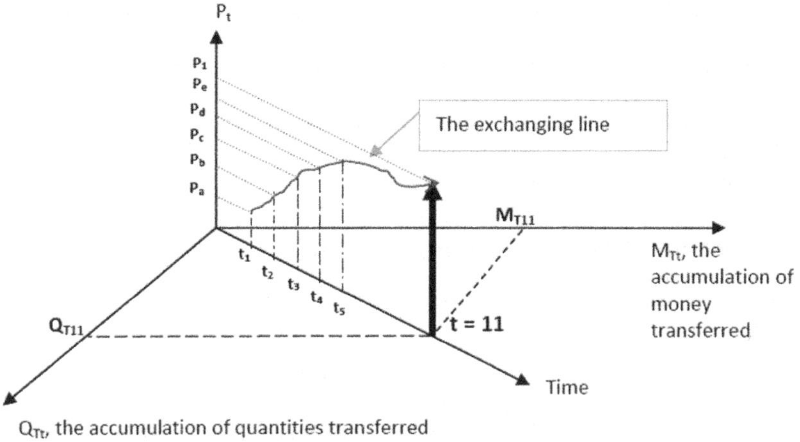

Figure 83: The price of a bottle of drinking water as the example used to present the exchanging line

In the above figure, the Exchanging Line used for the explanation of price in the real world is represented.

The fundamental concepts of new economics

Prices as data are the center of economy activities. In other words, prices are the universal set that consists of many subsets as interest, rent, wage, and profit, and others; so we should know about it their details first.

Average interest rate with time

Interest rate of loan is one of prices. Buyers are bank's customers who transfer his money called "interest rate" to bankers that are sellers and the money they borrowed from bankers called "capital" or "loan." Sellers are bankers who transfer their product as money called "capital" or "loan" to customers that are buyers. In other words, interest rates are the prices that capital owners (as sellers) tell their borrowers (as buyers) how much money the borrowers (as buyers) have to pay for the capital owners' hours (seller's hours) at the moment of exchanging.

For example, three borrowers, B, D and F, use three amounts of interest of three bankers, A, C, and E respectively. See Figure 84.

At t = 365, with products as money, 100,000+100,000+100,000 = 300,000 US dollars, sellers are three bankers, A, C, and E, transfer their loans to their customers as buyers. In addition, three amounts of money that are called "interest" as money per year, 10,000+11,000+12,000 = 33,000 dollars, are transferred from three borrowers B, D and F to three bankers respectively. Average interest rate of all borrowers is 11 % of 300,000 $ of loan per year (or 33,000 $ per 300,000$ of loan per year). This average interest rate consists of three values, P_B, P_D, and P_F, as follow:

(1) P_B = 10,000 $ per 100,000 $ per year
= 10 % of 100,000 $ per year

(2) P_B+P_D = 21,000 $ per 200,000 $ per year
= 10.5 % of 100,000 $ per year

(3) P_D = (2) - (1)
= 0.5 % of 100,000 $ per year

(4) $P_B+P_D+P_F$ = 33,000 \$ per 300,000 \$ per year

= 11 % of 100,000 \$ per year

(5) P_F = (4)-(2)

= 0.5 % of 100,000 \$ per year

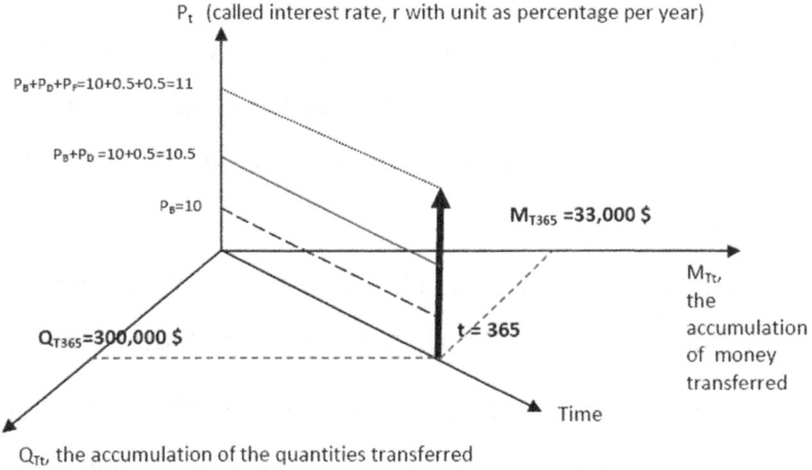

Figure 84: Average interest rate with time

The figure shows the new price model in economics. One of the prices in the real world represented in the figure is the interest rate. In the model, money owners tell their borrowers how much money they have to pay for the time spent while using money of the owners. This is the fundamental concept of economics in the past and future economics.

Average rent with time

Rent is one of prices as data that owners of properties (such as landlords and owners of cars) tell tenants (renters) how much money the tenants have to pay for time of using the properties of owners. For example, buyers are tenants who transfer their money called "rent" to landlords. Sellers are property owners who give their hours using their products to tenants. The properties are time of using of things as follow: a room, land, building, car, television set, etc. In other words, rents are the prices that owners of property (as sellers) tell their renters (as buyers) how much money the renters (as buyers) have to pay for the hours (seller's hours) at the moment of exchanging.

For example, three renters, H, J and L, use three rooms of three landlords, G, I, and K respectively. See Figure 85.

At t = 365, the renter H (or a buyer) pay 6,000 dollars (500 dollars per month) to his landlord G. Similarly, the two renters H and J pay 18,000 $ (6,000+12,000 per year or 500+1,000 dollars per month respectively). In addition, the three renters H, J, and L pay 36,000 $ (6,000+12,000+18,000 per year or 500+1,000+1,500 dollars per month respectively).

Average rent of all renters is 36,000 $ per 3 rooms per year or 12,000 $ per room per year. This average rent consists of three values, P_H, P_J, and P_L, as follow:

(1) P_H = 6,000 $ per room per year
= 500 $ per room per month x 12 months

(2) P_H+P_J = 18,000 $ per 2 rooms per year
= 1,500 per 2 rooms per month x 12 months

= 9,000 $ per room per year

= 750 per room per month x 12 months

(3) P_J = (2) - (1)

= 250 per room per month x 12 months

= 3,000 $ per room per year

(4) $P_H+P_J+P_L$ = 36,000 $ per 3 rooms per year

= 3,000 per 3 rooms per month x 12 months

= 1,000 per room per month x 12 months

(5) P_L = (4)-(2)

= 250 per room per month x 12 months

= 3,000 $ per room per year

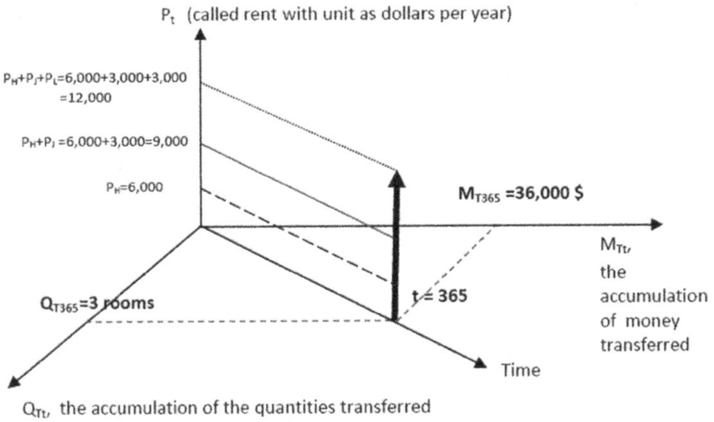

Figure 85: Average rent with time

The figure shows new price model in economics, which is represented in the real world with "rent." Property owners (such as landlords) tell their renters how much money they are required to pay for the time they use a property. Besides, this model forms one of the fundamental concepts of economics in the past and next economics.

Average wage with time

Wage is one of prices as data that employers tell employees who work how much money the employees will get for time of working for employers. Buyers are employers who transfer their money called "wage" to employees that are sellers. Sellers are employees who give (or transfer) their working hours to employers' works. The product in this case is working hours with experiences of employees. In other words, wages are the prices that employees (as sellers) tell their employers (as buyers) how much money the employers (as buyers) have to pay for employees' hours (seller's hours) at the moment of exchanging.

For example, three employees, M, O and Q, spend their time for three employers, N, P, and R respectively. See Figure 86. Average wage of all employees is 180,000 $ per 3 men per year or 60,000 $ per man per year. This average wage consists of three values, P_M, P_O, and P_Q, as follow:

(1) P_M = 36,000 $ per man per year

= 3,000 $ per man per month x 12 months

(2) P_M+P_O = 96,000 $ per 2 men per year

= 8,000 per 2 men per month x 12 months

= 48,000 $ per man per year

= 4,000 per man per month x 12 months

(3) P_O = (2) - (1)

= 1,000 per man per month x 12 months

= 12,000 $ per man per year

(4) $P_M+P_O+P_Q$ = 180,000 $ per 3 men per year

= 15,000 per 3 men per month x 12 months

= 60,000 $ per man per year

= 5,000 per man per month x 12 months

(5) P_Q = (4)-(2)

= 1,000 per man per month x 12 months

= 12,000 $ per man per year

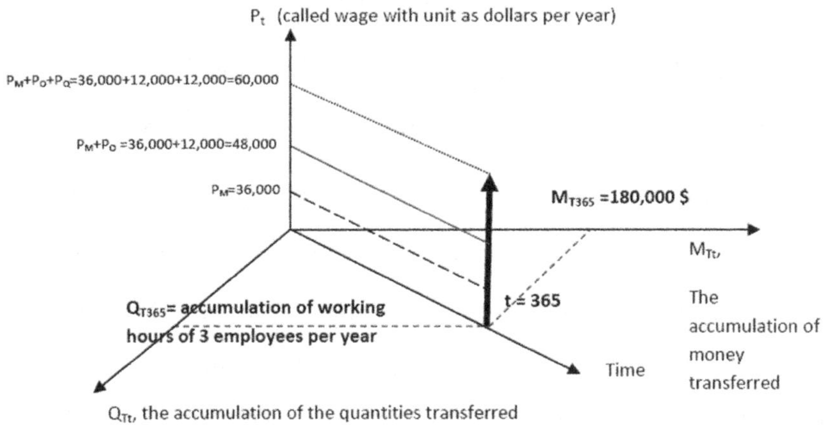

Figure 86: Average wage with time

In the figure, the new price model in economics is shown. The representative of the price indicated in the model as that of the real world is "wage." Force owners (such as employees or laborers) tell their customers (employers or capital owners respectively) how much money they have to be paid for the time they spend in the Force (such as brain force, body force).

Also, the model is considered to be a critical concept of economics in the past and for the future economics.

Average profit with time

Profit is one of the prices as data that entrepreneurs tell themselves how much money they have to pay for time of working their works.

Profits are the prices that entrepreneurs (as sellers) do not tell to their customers (as buyers) how much money the customers (as buyers) have to pay for entrepreneurs' hours (seller's hours) at the moment of exchanging.

For example, three entrepreneurs, A, B, and C, spend their time for management their business. See Figure 87.

At t = 365, the entrepreneur A get 120,000 dollars (10,000 dollars per month). Similarly, the two entrepreneurs A and B get 360,000 $ (120,000+240,000 per year or 10,000+20,000 dollars per month respectively). In addition, the three entrepreneurs A, B, and C get 720,000 $ (or 120,000 + 240,000 +360,000 per year or 10,000 + 20,000 + 30,000 dollars per month respectively).

Average profit of all entrepreneurs is 720,000 $ per 3 men per year or 240,000 $ per man per year. This average profit consists of three values, P_A, P_B, and P_C, as follow:

(1) P_A = 120,000 $ per man per year
= 10,000 $ per man per month x 12 months

(2) P_A+P_B = 360,000 \$ per 2 men per year

= 30,000 per 2 men per month x 12 months

= 180,000 \$ per man per year

= 15,000 per man per month x 12 months

(3) P_B = (2) - (1)

= 5,000 per man per month x 12 months

= 60,000 \$ per man per year

(4) $P_A+P_B+P_C$ = 720,000 \$ per 3 men per year

= 60,000 per 3 men per month x 12 months

= 240,000 \$ per man per year

= 20,000 per man per month x 12 months

(5) P_C = (4)-(2)

= 5,000 per man per month x 12 months

= 60,000 \$ per man per year

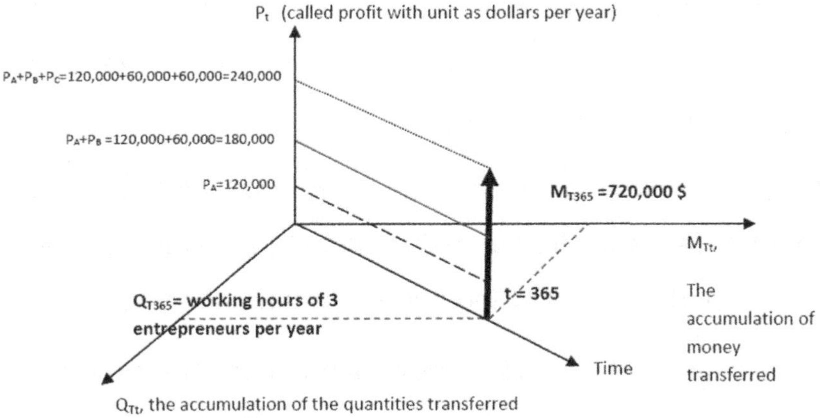

Figure 87: Average profit with time

In this figure, the new price model in economics is shown; but in this case, the price in the real world is "profit". The scenario in this model is where the product owners (such as entrepreneurs) tell (in silent) their customers (such as buyers) how much money they have to pay for time of using a product (tangible product or intangible products).

Average tax with Time

Tax is one of prices as data that government tells people called taxpayers how much money the people have to pay when they are buying goods and services of their sellers in the country. Taxpayers are paying their money called "tax" for getting goods or services legally.

Government tries to control people doing four main economy activities and all economy activities being run smoothly without quarrel. Taxpayers can be both buyers and sellers. Sometimes the sellers try to transfer their tax burden to buyers who are buying sellers' goods if they can.

There are many kinds of taxes in the real world such as Ad volarem tax, specific tax, value added tax, export tax, import tax, property tax, travelling tax and inheritance tax.

For example, at t = 365, three groups of people in the country pay average tax 33,000,000 dollars to their government (or 10,000,000+11,000,000+12,000,000 $). See Figure 88.

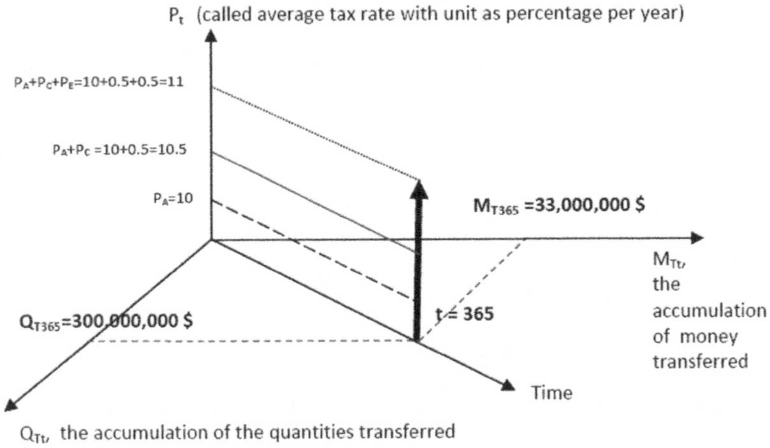

Figure 88: Average tax with time

The new price model in economics in the above figure is shown in "tax", to represent the prices in the real world. In this case, the government tells their people how much money they have to pay for time they use products of the country.

Average tax rate of all of people in this country is 11 % of values of things that government use to calculate their tax 300,000,000 $ of per year (or 33,000,000 $ per 300,000,000$ of values of things per year). This average tax rate consists of three kinds of percentage of average tax, P_A, P_C, and P_E, as follow:

(1) P_A = 10,000,000 $ per 100,000,000 $ per year
= 10 % of 100,000,000 $ per year

(2) P_A+P_C = 21,000,000 $ per 200,000,000 $ per year
= 10.5 % of 100,000,000 $ per year

(3) P_C = (2) - (1)

= 0.5 % of 100,000,000 $ per year

(4) $P_A+P_C+P_E$ = 33,000,000 $ per 300,000,000 $ per year

= 11 % of 100,000,000 $ per year

(5) P_E = (4)-(2)

= 0.5 % of 100,000,000 $ per year

The changing of Exchanging line

Exchanging line with time can show the weak point of the interest Keynesian's model. Classical economics indicated that the money market and the product market separated from each other, Keynesian economics indicated that the two markets linking to each other with interest. With the Keynesian economic theory, the interest rate comes from money supply and demand for money. If money supply increases, but demand for money in money market is constant with ceteris paribus; as the result, the interest with equilibrium following Keynesian theory will decrease.

However, the new price model indicates the two crucial markets linking to each other with the price as the data while buyers and sellers are exchanging their money with their goods, $P_t = M_{Tt}/Q_{Tt}$. The interest is one of the costs of producers, wholesalers, and retailers. The increasing of the interest rate of banks in the real world markets is the cause, which can make the increasing of prices.

The new price model shows that the open system of economy has a large group of variables. With characteristic of time, the time dimension is the link, which links all dimensions together. Nothing is permanent. Everything has been changing all the time. In addition, many present economy phenomena especially economy crises with the same causes will not occur repeatedly.

Exchanging line with time: the effect of an exogenous variable of neoclassical economics model

The new model shows that exogenous variables of present economists studying neoclassical economics model can be added in the model for studying. For example, government in a country that orders by law to all owners of the factories including the drinking water factory in urban area has to move to rural areas. The policy makes the higher costs of the owners of the factories including factories producing the plastic bottles, the covers, the stickers, the water in the bottles, the wage and else; as the result, these prices P_a, P_b, P_c, P_d, P_e, ... and P_1 change to P'_a, P'_b, P'_c, P'_d, P'_e, ... and P'_1 at ts = 12, 13, 14, 15, 16, ...22 respectively. See Figure 89 (Boonsri, 2014).

The examples of **political policies** of a country were discussed as follow:

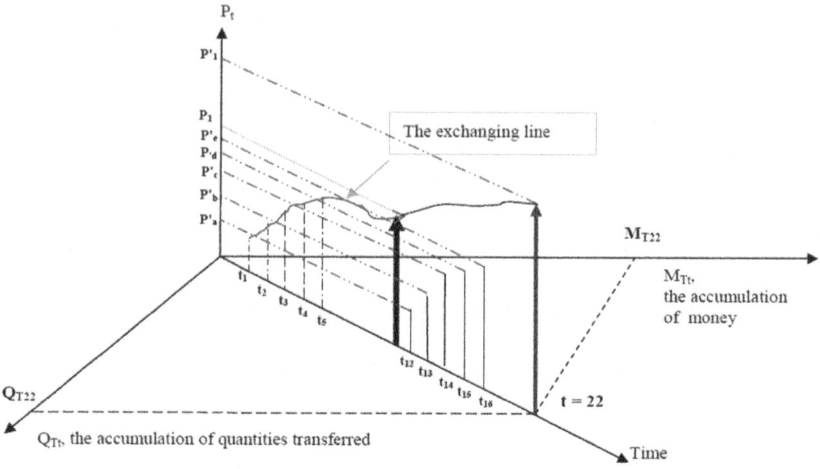

Figure 89: The exogenous variables of neoclassical economics explained in the new price model

This new price model in economics shows the exogenous variables that have never been studied in the economic models of any economy by any economist. The important variables in this model include political policy, religion, custom, culture, war, immigration, chaos, ethnic groups, and the greenhouse effect. All these can be used in the new price model to explain the effect toward the price in the real world.

In the future, economists will able to point out the different directions of development their countries for policy makers to reach the wealth of nations. The dynamics paradigm of the economics will be the new trend of the world. The economy crises that occurred in the past may be predicted the way to protect the same problem in the future. The next economy models will help economists in each country to see more clearly with wide and deep views.

For example, economists will understand more clearly what the weak points in economy in the past. Living together with friendship will be more beneficial because of reducing government expenditure on military and increasing national income. An economist will develop himself (or herself) to be an expert in the field that s/he is interesting in because the details of each market of each country are unique. The poor will decrease because the poor who can take care themselves more can produce higher economy values to their societies.

The rich will know that they can live more happily if they give the positive values to their societies. The linkage between microeconomics and macroeconomics that no economists including Alfred Marshall can do will be presented.

The comparison between Marshall's price theory and the new price theory was presented in the Table 7.

Table 7: The comparison between Alfred Marshall's price theory and the new price model

Items	Alfred Marshall's price theory	New price model
Methodology to create the theory	Deductive method	Inductive method

Relevance with Locke	Agrees with the Locke's concept: "price is a demand and supply theory"	Agrees with the Locke's concept: "price is a demand and supply theory"
Demand and supply sides	In price theory, there are the quantity demanded in the demand side and the quantity supplied in the supply side	In price theory, there are the money transferred in the demand side and the quantity supplied in the supply side
Statics/ dynamics	Comparative statics	Dynamics
Economic study	Normative economics	Positive economics
Curve	The demand and the supply curves are imaginary; it is not in the real world.	The exchanging line can be used to explain economy phenomena in the real world.
About time	Marshall was not interesting in time.	Time is the crucial link of this study.
Time and Linkage	The linkage among economy, politics, and society is not explained in the model	The time dimension is the link among economy, politics, and society in the model
Linkage	Microeconomics cannot link to macroeconomics	Microeconomics can link to macroeconomics

Model and assumptions	All models have the assumptions about exogenous variables	The model has no assumptions about exogenous variables.
Price	Price is not data	Price is the data of sellers telling buyers
Wage	Wage is not the data	Wage is the data of employees telling buyers, employers.
Rent	Rent is not the data	Rent is the data of landlords of places telling buyers, renters.
Interest	Interest is not the data	Interest is the data of owners of money telling buyers, borrowers.
Profit	Profit is not the data	Profit is the data telling nothing to buyers, customers.
Natural resources	Natural resources are public goods	Natural resources are public properties
Cost	Wage, rent, interest, profit	Wage, rent, interest
Revenue	In microeconomics	In micro- and macroeconomics
Break-even point	TR = TC = normal profit	TR = TC = break-even point
Exogenous variables	Exogenous variables are not studied.	Exogenous variables can be studied in the model

		= endogenous variables
Religions	Exogenous variable	Endogenous variables
Politics policies*	Exogenous variable	Endogenous variables
War	Exogenous variable	Endogenous variables
Immigration	Exogenous variable	Endogenous variables
Bargaining power	Exogenous variable	Endogenous variables
Culture	Exogenous variable	Endogenous variables
Education	Exogenous variable	Endogenous variables
Health	Exogenous variable	Endogenous variables
Consumption	Macroeconomics is different from microeconomics (demand)	Relevance between microeconomics and macroeconomics
Saving	Macroeconomics	Relevance between microeconomics and macroeconomics
Investment	Macroeconomics	Relevance between microeconomics and macroeconomics
Government expenditure	Macroeconomics	Relevance between microeconomics and macroeconomics

Export	Macroeconomics	Relevance between microeconomics and macroeconomics
Import	Macroeconomics	Relevance between microeconomics and macroeconomics
Public debts	Macroeconomics	Relevance between microeconomics and macroeconomics
Economy activities	Consumption, production, exchanging, distribution	Consumption, production, exchanging, distribution
Maximization	Profit of producers and utility of consumers	none
Market of the theory	Perfect competition market, monopoly, monopsony, monopolistic competition, and oligopoly.	Markets of all products and services in the real world. Each market has its own characteristic such as Toyota and BMW (market of cars).

* See how to use in Figure 89.

In Table 7, Marshall and I used different methods to create price theory. Marshall used deductive method, but I used inductive method. Marshall and I agreed with Locke's concept. Marshall used the quantity demanded in his demand side, but I used money transferred from buyers to sellers in the demand side.

Marshall and I have the quantity supplied in the supplied side. Comparative statics and normative economics are used in Marshall's price theory; but dynamics and positive economics are used in my price theory. Marshall created demand and supply curves that were imaginary that can be proved mathematically (but it was not discussed in this book), they are not in the real world; but I create Exchanging line that is in the real world. Marshall was not interesting in time dimension; but I was. Marshall had not any link among politics, economy, and society in his model; but I have the linkage among them in the model. Marshall could not link microeconomics and macroeconomics together; but I can link those levels.

Exogenous variables that are assumptions in present economy model were not studied; however, I do not have any assumptions on exogenous variables. As the result, the exogenous variables can be studied in my model. Marshall used perfect competition market in his model, but I use the real world market in my model. Price that is not data was made by demand and supply curves in Marshall's model; nevertheless, in my model, prices are data. Time links future sciences especially mathematics, physics, and economics together.

New views to solve economic problems in the real world

Keynes' theory with time is the linkage between microeconomics and macroeconomics (Boonsri, 2015). The Great Depression, worldwide economic depression, began in the United States, around September 4, 1929 with the fall in stock prices; Personal income, tax revenue, profits and prices dropped, international trade decreased by more than 50%, and unemployment rose to 25%. It varied across nations, started in 1930 and lasted until the late 1930s or middle 1940s. In some countries unemployment as high as 33%; it was the longest, deepest, and most widespread depression of the 20th century ("Great Depression in the United States", 2016).

In the 1930s, John M. Keynes, 1st Baron Keynes, disagreed with ideas of neoclassical economics; with free markets, full employment would be solved by invisible hands automatically in short or medium run term. Keynes advised governments should make loan for increasing employment with fiscal and monetary measures to solve the depressions. Following the outbreak of World War II, Keynes's ideas concerning economic policy were adopted by leading western economies. Keynes died in 1946; during the 1950s and 1960s, Keynesian economics helped many capitalist governments adopting its policy recommendations. In 1999, Time magazine included Keynes in their list of the 100 most important and influential people of the 20th century ("John Maynard Keynes", 2016; "John Maynard Keynes | British economist", 2016).

During 1966 and 1967, the models called FRB-MIT with Keynesian's theory with American monetary and fiscal policies were made for solving economy problems. Hundred equations made by leading economists working in universities and private sectors for representing sectors of production in all economy activities. Economists at that time believed that the result of the models could guide to effective implementation; so many policies following the result were made. One of the main objectives was to decrease unemployment rate, prices, and other variables in the models. However, before the models gave the results, economists had to ask government, which rate of inflation for decreasing rate of unemployment, could be accepted. The models would be one of the best policies. Planners could point how economic problems at that time would be solved completely.

With the development of computer technology, accurate data were provided for the computers to calculate; economists believed that government could control occurring of the business cycle at the suitable point. US government wanted to get unemployment rate less than four percentages with four percentages of inflation; however, in case of inflation rate rising to five or six percentages, unemployment would disappear. When policies from the models were implemented, the failure of the models made unwanted results. The rising of seven of unemployment rate and eight percentages of inflation rate occurred. The mistake came from the assumption that indicated there were opposite directions between the rates of employment and inflation; Keynes' theory could not explain unemployment and inflation in 1970s occurring simultaneously (Willes, 1981).

Economists cannot explain some limitation on prediction of Keynes' theory such as what condition that make Keynes' Theory work or what situation that Keynes' Theory could not explain.

Keynes explained about inflation with different views compared to classical economists because classical economic theory did not work well. In details, classical economists believed that inflation came from demand of people with constant supply, but Keynes explained that inflation came from demand and supply in economy system. Keynes divided demand into three levels; far from full, near full, and full employment in his demand pull inflation theory.

At far from full employment, if demand of people occurs; inflation will not occur. At near full employment, inflation will occur with low rate. At full employment, inflation will occur with high rate called hyperinflation. In his cost-push inflation theory, inflation came from workers by increasing wage and employers by increasing employers' profit. If workers got too high wages or the employers got too high profit for their business, inflation would occur.

In Keynes' inflation theory, increasing money supply was not explained directly but the money hiding in demand and supply indirectly.

About debt, Keynes advised Great Britain government should borrow money for using and activating economy activities during Great Depression; as the result, economy problems in Great Britain during Great Depression could be solved.

Right now, many economists still believe that one way to solve economy problem is that government has to use budget deficit. The budget deficit is an excess of expenditures over revenue in a given time period; as the result, many economists believe that if government wants his country to be rich; the government has to spend lot money for better economy.

Economic equation of Gross National Product at market price, GNP or C+I+G+X+M, shows if government spends more money on G; GNP will be higher; as the result, many governments in many country using enormous money to get better economy. In USA and Thailand, we can see megaprojects being made for society to get better economy. NASA is one of the megaprojects in USA. Nevertheless, the debts of many countries such as USA and Thailand are very high as well.

In addition, in macroeconomic theory, present economists believe that saving is one of leakages in economy system. If a country is trying to increase saving money of people, then whole national income at equilibrium will decrease and the new saving situation of people in the country compared to initial saving may be one of two ways: constant or decreasing rates of saving.

In my price model, there are no three levels of employment as earlier indicated in the Keynes' theory. In addition, there is relation among increasing price, inflation, money, money supply, and quantity of products and services. Time in the model can link microeconomics and macroeconomics together. Moreover, the model indicated that price is the link between money and product markets.

Is there something wrong on the way to solve economy problem? To solve this question, the new price model with the exchanging line and the effect of an exogenous variable on microeconomics and macroeconomics will be used to study the limitation on Keynes' theory.

The limitation of Keynes' theory with time

High inflation and high unemployment occurring simultaneously can be unexplained. There are three steps for indicating that time as the linkage between micro- and macroeconomics can explain how high inflation and high unemployment occurring simultaneously that Keynes' theory could not explain (Boonsri, 2015).

First, increasing money supply with demand for money as constant may make and may not make interest rates of saving and loan decrease following Keynes' theory.

In the real world economy, the Central bank determines the lower interest rates of loan because the Central bank believes that decreasing interest rate will make higher investment following the result of Keynesian' s theory. In addition, the Central bank believes that commercial banks will decrease interest rate following the Central's announcement. If the Central bank and commercial banks decrease the interest rates of loan, the Central bank believes that the cost of firms about loan will decrease which firms will increase their investment in economy system. However, there are no rules especially laws of government to force commercial banks if commercial banks do not decrease their interest rates following the Central bank. The commercial banks run business on money; they have to think of their money first. If they decrease the interest rate; it means that they will receive money from the firms who borrow money from them, they are decreasing their revenue.

The customers who borrow money from banks can be classified following the criteria of bargaining power: low, medium, and high. If commercial banks have higher bargaining power than the customers, they may not decrease interest rate of customer's loan.

If the commercial banks decrease the interest rate of loan, they decrease their revenue; as the result, they have to decrease interest rate on saving of other customers as well because interest rate on saving of other customers is the cost of banker's business. Decreasing interest rate of saving can make customers with "high bargaining power" withdraw their money from the bank account to other places for investment. If no any law to force the commercial banks, it is possible that commercial banks may or may not change the interest rates of loan and saving with the same rate of the Central bank's announcement.

The conclusion in this part is that in Keynes' theory, increasing money supply with demand for money as constant will make the equilibrium interest decrease. In other words, in Keynes' theory, money supply and demand for money are the causes, the decreasing interest rate is the effect, and increasing investment is the effect later.

However, with the Keynes' theory and the new price function and time, government by using the Central bank can announce the change in the interest rate without increasing money supply with demand for money as constant; as the result, commercial banks may and may not decrease interest rates of saving and loan following the Central bank of a country. In other words, when the Central bank changes interest rate following government policy, commercial banks may or may not change the rates following the announcing. In other words, interest rate of the Central bank does not change because of Keynes' theory, the change on money supply with constant demand for money.

Second, without increasing products and services in economy system, increasing money supply causes higher prices of many products, inflation as follow:

From the new price function, $P_t = f(M_{Tt}, Q_{Tt}) = \frac{M_{Tt}}{Q_{Tt}}$, there is the direct relation between money transferred from buyers to sellers and prices if quantities transferred from sellers to buyers as constant.

For example, at t = 1, a product transferred from buyers to sellers, Q_T, are 100 units of the product; and money transferred from sellers to buyers, M_T, is 100 units of money; as the result, P of the product is 1 unit of the money per unit of the product. At t = 2, the product is transferred from buyers to sellers as 100 units and money is transferred from sellers to buyers as 200 units. As the result, P of the product is two units of the money per unit of the product; this shows inflation will occur in the system when the money is increased with the constant of the product that is transferred from sellers to buyers.

This result of the new price model shows the same strong empirical evidence of a direct relation between money-supply growth and long-term price inflation, at least for rapid increases in the amount of money in the economy in the real world market. For example, Zimbabwe, which saw extremely rapid increases in its money supply, also saw extremely rapid increases in prices (hyperinflation) ("Hyperinflation", 2016).

Third, if the prices of four requisites and energy increase, then the cost of people's living will cause higher inflation in the society.

If the prices of four requisites and energy increase, then the cost of living of people that was represented by consumer price index, CPI, will cause higher inflation in the society. The higher prices will cause employees increasing their wages from employers, and will cause higher inflation as well. The higher wages may make some firms decrease their employees; this leads higher unemployment. The conclusion in this part is that the higher inflation and higher unemployment can occur simultaneously.

Wages is one of prices. Buyers are owners of firms, employers, who transfer their money called "wage" to employees for employees' working hours. Sellers are employees who transfer their product as the time for working to employers that are buyers.

Because of employer's profit, if an employer has to pay too high wages for employees' working hour, then the unemployed situation will occur. Normally, the employer does not want to pay wage for his or her employees who cannot make money high enough for him or her. In addition, the employer does not want to pay high wage as well, if s/he can find a new employee who receives lower wage with same quality of works.

While Keynes's theory cannot explain how high unemployment and high inflation occur simultaneously, the time as the linkage between microeconomics and macroeconomics can do; this shows limitation of Keynes' theory with normative economics; in contrast, this shows the result of the new price function with positive economics.

The condition of Keynes' theory works well

Comparison among three conditions for indicating that while government is trying to increase money supply in economy system, they have to increase products or services of many people for avoiding inflation (Boonsri, 2014).

First, for avoiding inflation, while they are trying to increase money supply in economy system, government has to increase products or services of many people such as employing people to get public works: constructing interstate highways or growing plants for increasing forest areas. In addition, government has to support people more factors of their production. However, government should not increase products and services of the rich who may hold a lot of money in his/her own hands without widely money distribution. See Figure 90.

Second, increasing money supply with mega project such as the project of Space Station without increasing products or services of many people in economy system makes a little group of people get richer and makes wider income gap between the rich and the poor. See Figure 90.

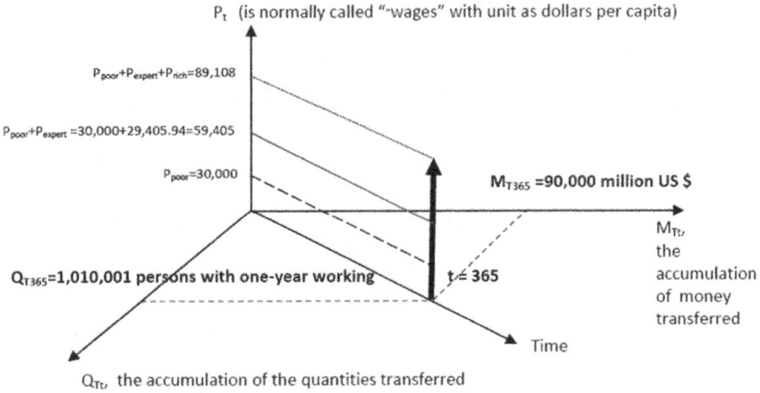

Figure 90: The transferring money to different kinds of people makes different results for the wealth of nation.

In the figure, the new price model is used to point out the limitation of Keynes's theory to solve economy problems in the real world. For example, while government is trying to increase money supply, 90,000 million dollars, in economy system; government chooses three kinds of people: (1) 1,000,000 poor people; (2) 10,000 experts who run mega project such as

Space station; and (3) the only one rich man/woman who runs his/her own business.

Third, government budget deficit without increasing products or services of many people makes higher debts that will be the burden of the society in the long run for paying back the debts, capital and interest rate.

At t = 365, if 1,000,000 poor people who are sellers that transfer their products, one-year working for public works to government such as constructing interstate highways or growing plants for increasing forest areas; the government that is a buyer will transfer the last amounts of money called "wages" to all poor people. With total wages, 30,000 million US dollars, average wage is 30,000 dollars per capita per year. If the average value of marginal propensity to consume, MPC, of the poor people is 0.90; this means that large amounts of money, 30,000 x 0.9 = 27,000 million dollars will be exchanged for consumption in the economy system.

At t = 365, if 10,000 experts who are sellers that transfer their products, one-year working on mega project to government; the government, a buyer, will transfer money called "wages" to the experts. With the amounts of the whole wages, 30,000 million dollars, an expert will get 1 million dollars per capita in average. If the average value of marginal propensity to consume, MPC, of the experts is 0.50; this means that the amount of money, 30,000 x 0.5 = 15,000 million dollars will be exchanged for consumption in the economy system.

At t = 365, if the rich who is a seller that transfers his/her product, one-year working hours to government; the government will transfer money called "wage" with total wage, 30,000 million dollars (or 30,000 million dollars per capita). Normally, the rich wants his/her profit maximization; so s/he will offer minimum wages for his/her employees to work his/her business because of higher wages making lower profit. If average value of marginal propensity to consume, MPC, of the rich and the employees is 0.30; this means that the amounts of money, 30,000 x 0.3 = 9,000 million dollars will be being exchanged for consumption in the economy system. This way will increase the wider gap between the rich and the poor as well.

While government is trying to increase money supply in economy system; if there is no investment expenditure, then government has to increase products or services of many people. See Figure 90.

Keynes' theory cannot explain how and when their theory works; this shows limitation of Keynes' theory with normative economics. However, time as the linkage between microeconomics and macroeconomics can explain how and when Keynes' theory works with positive economics.

In other words, in Figure 90, government transferred the wages to three kinds of people. Average wages of 1 million people, 1,010,000 people, and 1,010,001 people are 30,000, 59,405, and 89,108 dollars per capita per year respectively (Boonsri, 2015). See Figure 90.

With my new price model, inflation will occur at three levels compared to Keynes' theory, if we increase money supply without increasing products in economy system. With my model, mega projects will be beneficial for some group of people who closely concern, not all people with wide money distribution.

Mainstream economists always say it is good if government increases government expenditure or has mega projects because huge amount of money will be used in economy system. For example, with the project of NASA in USA and flood protection in Thailand, big amount of money will flow to owners of private firms that have right for building infrastructure.

Small amount of money will not flow to big amount of people. In other words, most people in my country will not be beneficial because a small group of people in the countries runs the projects. If government wants to get better economy, government has to help most people having their works directly.

In addition, government has to teach people how to fish, not give them fishes. Government has to increase the factors of working of most people. For example, if people are farmers, government has to support seeds to them for cultivating and increasing higher velocity of their supply chains to end users.

If government helps people "standing stronger on their feet or teaching people how to fish," better economy will occur (like the project of King Rama IX, the great). The project of "Food Stamp" in L.A. is the good example that government gives fishes to people.

If government uses budget deficit for better economy but the loan is not used for helping most poor people having their jobs or for "standing stronger on their feet or teaching people how to fish," this way will cause higher debt of the country. For example, if government creates the project of increasing areas of forests or constructing infrastructure of society. Money has to be used directly to the poor for supporting their "how to fish." In other words, increasing money in economy system without increasing products and services will cause bad, not good results for country.

Other ways to increase money supply in the economy system

Other ways to increase the amount of money in economy system are government decreases taxes and commercial banks increase their loan for firms. Decreasing taxes will increase money in economy system with the condition that taxes has to be used for the poor with high rate of tax decreasing, not for the rich because when the poor gets more money in their pockets, they will use their money for more consumption; as the result, the better economy will occur (Boonsri, 2015).

Commercial banks increase loan for firms to run business, this way will support better economy through entrepreneurs and employers, if government helps the activities with supply chains run smoothly.

Export is the other way to increase money in economy system; however, there is a weak point in this method. If government supports a small group of people to be exporters, many people will not be beneficial from export. Exporters will be the rich but most of people are still the poor because most exporters will try to buy cheap goods to sell higher prices.

Increasing money in economy system does not guarantee to make most of people be richer; these events used to occur in the past of Europeans. In Europe from the 16th to the 18th century, economy theory and practice was Mercantilism. Wealth of nation came from accumulating monetary reserves through a positive balance of trade, especially of finished goods ("Mercantilism", 2016; "mercantilism | economics", 2016). Only small group of people was richer such as some noblemen and a king or a queen who were trading with their overseas colonies; however, if government did not help most of people as a new group of exporters, then most benefit from export would belong to "big exporters."

Other ways to increase money in economy system

Many countries in this world believe that investment from foreign countries will make better economy as well. With my model, if there are many new foreign factories in a country making high pollution, economy in the country will not be better with long run.

Small group of people will get directly beneficial. With higher pollution, in long run, the investment will make higher economy problems. Government has to use taxes of people to solve pollution problems. The owners of the factories will get beneficial but the rest people won't. The example shows there is some big hole of leakage occurring when government tries to increase investment in his country with foreigners.

Increasing amount of money in stock market is one of the ways to increase money in economy system. In Keynes' theory, there are three reasons of people who want to have money in their pockets called "demand for money" in economy system: (1) for buying food, clothes, and houses in every day, (2) for emergency cases such as getting sick or accidents, (3) for speculation in a stock market. Increasing money in the stock market makes demand for money increase. If money supply in economy system is constant and demand for money increases; as the result, interest rate will increase. When interest rate increases, investment in economy system will decrease. National income will decrease next; these came from Keynes' theory (Boonsri, 2015).

With my new price model, increasing amount of money in stock market will be beneficial for people who have to have their saving money in long run. Some people who borrow money to trade in the stock market will be higher risk of losing money because of the decreasing rapidly in the stock market index in short or long run makes them be forced to sell their stocks and getting “true” losing. In addition, the people will make enormous economy problem if they use money of other people including of banks as same as the occurring the Great Depression in the past because of this activities playing with enormous amount of money in economy system. Many poor people will not get any directly beneficial.

Real world economic theories

The core in economics is price theory, so the real world economic theory starts at this point. The new price model can be used to explain production in the real world. Production in the real world following neoclassical economics theory can be divided into three economy systems: Capitalism, mixed, and communists. Right now, most of countries in the real world are mixed system. Government and private sector are pushing economy activities, and both of them are the owners of their products and services, of their land, of their labor forces, and of their capital. Production in the real world can be divided into 5 steps following development of societies: (1) production of primitive society, (2) production of advanced agricultural society, (3) production of industrial society, (4) production of information and technology society, (5) production of wisdom society.

Production of primitive society is agricultural products. Farmers always grow plants for food with depending on nature such as weather, soil, water, animal force, and seeds. Sometimes, farmers try to use "modern" ways to increase their yield especially using chemical fertilize, biocide; however, when they grow monotype of plant with using chemical fertilizer and biocide for a long time, finally, lack of good quality of the soil and the water to produce new products took place. Normally, farmers without any new information will be poorer and poorer. (This case was found in rural areas of Thailand.) And the nation was underdevelopment country. When people in the country are poorer, other serious problems occur continuously especially healthy, and education. Sometimes, these serious problems are called "Poor cycle." (or poor, illness, low education.)

Production of advanced agricultural society

If government of the poor country wants to break down the cycle, the government has to solve this cycle by using advanced agricultural society. The objective of development has to emphasize to support people's opportunity to do their careers that they want or their choices. Government in this society, for example, solves problems about four factors for living as follow: food, dresses, houses, and medicine. For example, in short run by giving things (giving fishes) but in long run by support people's opportunity to do their careers (helping people how to fish) have to be done. In long run, for example, the quality of soil such as solving acid soil, growing vetiver grass, the quality and the quantity of water such as reforesting, making artificial rain, making check dams, weirs, dams, big and small reservoirs, communities ponds, and smallest pond in their farms individually, the quality and the quantity of plants that fit for the areas such as plant breeding, hybrid varieties, selection, varieties test, for food and trade, the quality and the quantity of roads for people to transportation, the quality and the quantity of people's education for their carriers such as establishing of demonstrating farms, the quality and quantity of people's marketing such as in short run, government has to set an organization as marketing services department for sending farmers' products to end users or buyers, and in long run, government has to teach farmers how to run their own business. The weak points of government in this society are the honesty of civil officers because if corruption occurs, it will break the speed of development.

The successful case study: projects of H.M. the King Rama IX, the king of Thailand during 1927 to 2016

Large crop areas in Thailand were done and did not make the wealth of nation enough. When large areas of corn were grown and insects such as Bombay Locust or diseases of plants damaged the fields, it was hard to protect large fields in very short time. For solving the problems, rich country such as U.S.A. use insecticide, side effect of the solution make another problem occur. For example, first, water with the biocide makes many animals living in the water mutants, and second, insect that did not be killed by insecticide increase natural immunity for survival of the next generation. Third, insecticide always makes bad healthy for people who touch it directly and indirectly in the long term. Fourth, farmers with large farming have to borrow money from banks, money from banks makes the farmers higher cost every second. Fifth, farmers have to wait for rain to cultivate, when lack of rain or too much rain because of abnormal weather from greenhouse effect, then the farmers will lose their all or a part of products. The farmers' debts make farmers lose their lands. Sixth, with families having many children, large farms are always divided into smaller and smaller size. Finally, farmers in the next generation have not enough area; they have to increase their costs by renting another land. This makes higher cost and risk of cultivation. Integrating farming can solve the problems in Thailand. Seventh, merchandizers got richer and richer because of selling chemical fertilizer, biocide, getting high interest that farmers borrow their money, getting larger land from their loans in the case of getting lost of farmers.

In Thailand, H.M. the King Rama IX, the great, established the foundation called "The Chaipattana Foundation (the victory of development)" to do more than 4,700 projects that were called projects initiated by His Majesty King or The Royal-initiated. The NGO working with Office of the Development Projects Board controlled directly by Prime Minister of Thailand to develop the country for Thai people for 70 years. Most of people and most economists in Thailand realized that the projects are a part of the wealth of Thai nation. At the beginning of the reign of H.M. the King Rama IX, Thailand was one of the underdevelopment countries in South East Asia that has its boundary connecting to Laos, Cambodia, Myanmar, and Malaysia. At the end of his reign, Thailand is one of the developing countries with high income. In this topic, the new price model will be used to explain how the projects have been benefit for Thai people until now (2016). **Third,** government budget deficit without increasing products or services of many people makes higher debts that will be the burden of the society in the long run for paying back the debts, capital and interest rate.

The whole picture of the projects is the new way to solve economy problems in Thailand. The objective of the projects concerns three words: (1) “exploding from within a society”, (2) our loss is our gain, and (3) “help and teach people how to fish”

The economy system of Thailand is mixed system. Many Thai people are working in agricultural sector. Most of the poor are in the sector. The crucial factors for agriculture are water, soil, and labor force. The quality of soil of many rural areas was not good enough for farmers to grow plants for food. The causes of the bad quality of the soil in the areas are different from each other. Some came from the nature, some came from men by cutting too much trees in forests for cultivation that made finally the ground not be able to grow any plants.

In the web page of ministry of agriculture and cooperatives in Thailand, this detail was written to explain what the king did projects: “His majesty the king in the capacities of a thinker, an inventor, and an experimenter, have been striving to seek ways of and approaches to the resolution of problems be leaguing Thai farmers, ranging from soil problems by increasing fertility for sandy loam, and increasing moisture for hard setting soil, and laterite soil, use of irrigation systems to reduce salinity for saline soil, and solving the problem of acid soil by affecting the soil compositions vicariously through a process modelled after the natural process, whereby soil acidity is gradually reduced resulting in higher levels of soil fertility” ("History", 2016). I called the projects “projects concerning soil.”

All projects initiated by the king working with Office of the Development Projects Board, ODPB, can be divided into four categories of solving with 10 main kinds of work: (1) solving the poverty problems (economy problems), (2) solving the educational problems, (3) solving the healthy problems, and (4) solving emergency cases. The first category solving poverty problem consists of three main works: (1.1) projects concerning soil, (1.2) projects concerning water, (1.3) projects concerning food plants and fishes, and (1.4) projects concerning skills for working. The second category solving educational problems consists of two main works: (2.1) projects concerning education, (2.2) projects concerning communication. The third category solving healthy problems consists of two main works: (3.1) projects concerning healthy, (3.2) projects concerning better life and happy mind, and (3.3) projects concerning transportation. The fourth category solving emergency problems consists of one main works: (4.1) projects concerning natural disasters relief and recovery.

The only one objective of all four categories is helping people to "fish" by using of their specialization. This can be referred to as the "people's hands." Farmers and their livelihoods are held at the center of the Chaipattana Foundation's development work. The Foundation aims for improving their social welfare and capacity to become self-reliant. The Foundation promotes His Majesty the King's principle of sustainable development. Rather than immediate results, it invests in projects that will yield long-term results and help the people to help themselves. The Foundation always works with the people whose lives will be affected by its implementation of projects. Their opinions will be listened and incorporated in the decision-making process, and they are encouraged to choose their own direction of development ("Missions and Operating Principles", 2016).

The projects concerning water were done at many levels. The project makes artificial rain to rain in areas lacking water. The way of water flowing may make new wet forests by making check dams or weirs. The water will move from the high to the low. Before the water flows to the land for cultivation, it may flow to dams for using water in growing seasons and making electricity. The water from the dams will be flown to smaller places such as big and small reservoirs. From the reservoirs, the water will be separated to each farmer. Farmers will make own smallest ponds to grow whatever they want to eat (and eat whatever they grow and feed).

With integrating farming, the area of a farmer will be used for at least four objectives: (1) 30% of the area is used for their own pond. In the pond, fishes will be fed for food and trade, and the water in the reservoir will be used for growing rice, vegetables, and fruits. Above the pond, chicken in the poultry may be fed for food and eggs, their chicken dung will be food of plankton, fishes eat plankton for their food. Second, 30% of the area is used for growing rice, and the last 30% of land is used for growing vegetables and fruits including feeding fogs for food. The rest 10% is used for farmer's houses. This flexible ratio may be adjusted for fitting with farmers in the different areas.

With this method called "New Theory", farmers will decrease their expenditure for living. They eat whatever they grow and feed, and they grow and feed whatever they want to eat. Their excess products will be sold or exchanged with their neighbors. The rest of products of farmers and neighbors that establish their cooperatives will be sold in farther areas. They will have income every day. The prices of their products will not be too low at the same time. Sometimes, the price of rice may be low, but the other prices of products such as eggs, mushrooms, chicken, and fogs will not be low.

In the case of flood, large amount of water will have to be managed by using flood ways to flow the water and special areas called "monkey's cheek" for keeping water. In case of polluted water in urban areas, waste water treatment by using aerators moving with very low energy and using water plants will be done. After many kinds of agricultural products are produced, farmers will eat whatever they grow, and grow whatever they eat. The three projects initiated by the king about water, plants, and transportation can be explained with new price model as follow:

At t = 1, before king's project took place, a farmer had his average income per year P_1Q_1 or M_1. See Figure 91. He had very low income and wanted to sell his field but he could not find a buyer. At t = 2, after the first king's project about water had completed, he got his higher average income per year P_2Q_2 or M_2 ($M_2>M_1$). If he wanted to sell his field, then he found two men living near village who wanted to buy because of enough water for cultivation. At t =3, after the second king's project about plants such as strawberry and coffee had completed, he got his higher average income per year $P_3.Q_3$ or M_3 ($M_3>M_2$) because of many agricultural products sold by members of King's Foundation. He got his full money without paying anything's for marketing. If he wanted to sell his field, then he found twenty men near his village who wanted to buy because of fruits making good benefit.

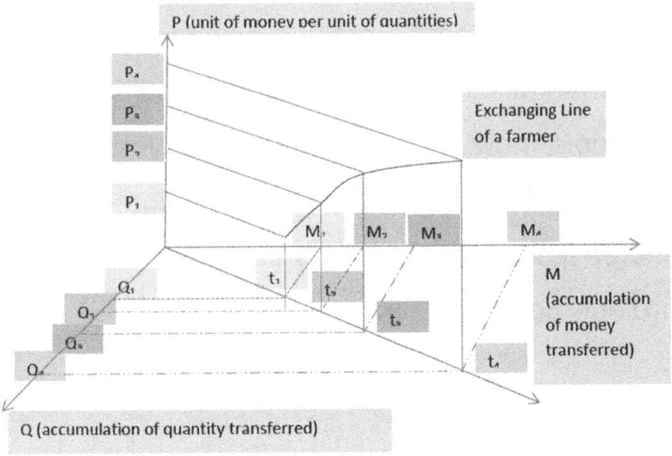

Figure 91: : The first to fourth incomes per year of a farmer are $P_1.Q_1$, $P_2.Q_2$, $P_3.Q_3$, and $P_4.Q_4$ respectively. In addition, $M_1.< M_2 < M_3 <M_4$.

In this new price model, what is pointed out is why the work of the king of Thailand can develop economy of Thailand successfully. Besides, the model can be used to investigate the success of a government's policy towards the growth of an economy.

At t =4, after the third king's project about road, school of his children, and hospital had completed, he got his higher average income per year $P_4.Q_4$ or M_4 ($M_4 > M_3$) because of many tourists visited his filed and bought his agricultural products in front of his field. He did not want to sell his field although the rich offered him big money. This real case occurred at the north province of Thailand in the past. See Figure 91.

After king's third project, government wanted to follow king's project to increase the income per year of 1,000 families of farmers. The government created a new project called "selling at internet market." The project still has the same objective, helping people how to "fish", after one-year working, the government got the result that: average income of people per capita increases ten percentages compared to the beginning of the project. See Figure 92.

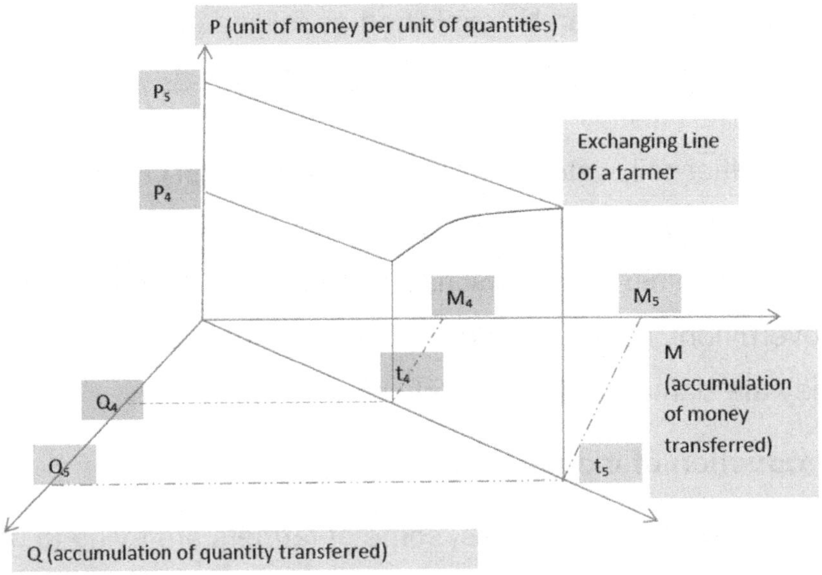

Figure 92: The increasing of 10% of the farmer after his joining of government's project at t_5.

The new price model was used to point out the prediction of Government's projects and their potential impacts in terms of the how they would successfully develop the economy of Thailand or not. Conclusively, the solution of economy problems in the real world can be predicted by using the new price model.

He got his higher average income per year $P_5.Q_5$ or M_5 ($M_5>M_4$) because his products were developed as community products sold by himself in front of his field and cooperatives by using internet. In other words, the rate of change of the farmer's income can be explained as follow: At P_5, the rate of income growth of the farmer = 10% = $\left(\frac{P_5Q_5 - P_4Q_4}{P_4Q_4}\right).100$

Notice that soil, water, and forest are three most crucial natural resources for agricultural production. Polluted water in canals, rivers, or other reservoirs has to be controlled seriously by government. If people make problems on the public resources, they are damaging the property of all in the country.

Production of industrial society

In the developing countries, some of farmers are trying to use agricultural machinery from aboard, this leads higher farmers' cost of agriculture with lower income if products they get are constant. The situation of modern without development still takes place continuously because people and the government do not make their own ways to support their production. Merchandizers and importers are still rich by promoting and selling their agricultural machinery. They always teach and transfer technology of maintenance with high costs for buyers' countries, not highest technology that they know. Technology from aboard may help people to produce their products faster but people still depend on technology without the "root" of the technology that people have to pay very high costs.

If government wants to develop the country to industrial society, the government has to develop people's education to have their own "root" of technology of agricultural machinery and equipment to get higher income. For example, without the root of technology, cutting of solid laser, making high-end cars, planes, submarines, making high-end of jet engines could not be invented. However, countries with high-end technology are supposed to be supported by inventors of agriproduct machinery because we need food, not mobile phone for living.

The leading countries such as the United States of America, Great Britain, Germany, and Japan had two kinds of technology before they reached to higher levels of technology. They always buy cheap raw materials and sell expensive agriproduct processing.

The objective of development emphasize to support people's opportunity to do their careers that they want or their choices (the only one direction of economy development). Government in this society, for example, still solves problems about four factors for living as follow: food, dresses, houses, and medicine. Normally, tropical countries will not have many problems about clothes but countries near the poles of the earth have.

For example, in short run by using their places to be temporary markets to help farmers selling their products to end users or buyers without merchandizers. The government may use a reward for their civil officers who work well. The government may use reduction of tax for people who set temporary market in their communities. In long run, the government still supports people's opportunity to do their careers (helping people to fish). For example, developing on the quality and quantity of people's marketing ways such as setting strong cooperatives, linking cooperative together, and making internet infrastructure for farmers to sell their products to other farmer's networks and people in different areas.

In addition, government has to teach farmers how to run their own business in higher levels. The government develops on quality control of production, packaging, standard of selling to buyers by using their civil officers. In addition, the government develops agricultural machinery (or farm machinery and equipment) to help farmers' production by using low-end technology first and middle-end and high-end technology in the next steps such as biodiesel, factories that farmers are a group of owners to produce tools, instruments, machines for agricultural, and factories for agriproduct processing. Moreover, the government has to develop the skills and education of people for their high-tech carriers.

If harvesters, for example, are made in the countries, and if they have 100 machine parts, that means the growth of 101 industries being made as well. The agricultural machinery may be export for higher income of stakeholders including farmers and countries' income. Farmers will buy their 100 parts of the product for maintenance as well. After industries of agricultural machinery and equipment are taken place, the other industrial products are the next step of production. The most benefit in the long run of the society is the "real technology for development" having been established, not modern without development anymore.

The weak points of government in this society are the honesty of civil officers because if corruption occurs, it will break the speed of development. The government may increase the income of people by establishing the projects of people's houses in prices without profit or without merchandizers. The government is supposed to support private sector to do the other businesses except business of four factors such as food, houses, dresses, and medicine. The projects will create new hundred industries concerning to them.

The successful case study: His Majesty the King's Initiative on Renewable Energy

For the past 20 years the world has been faced with an energy crisis. In 2000, the cost of the imported oil reached 258,172 million baht, with up to 12,133 million baht paid for diesel oil alone, which is an important factor for the industrial, agricultural and transportation sectors. This led Thailand to a trade deficit and raised the prices of consumer products, which in turn had caused people to suffer.

His Majesty the King sought methods for renewable energy production to benefit the people and national development at large. In the beginning, he initiated a study on the use of wind, water and sunlight energy to be used at royal development projects. Later, he granted another initiative to study the use of agricultural products to replace oil. This study concerned production of gasohol to mix with low octane benzene to increase the octane value. His Majesty granted an initiative to conduct a study on the use of palm oil to replace diesel oil used by machines. This led to the establishment of the Royally-Initiated Project for Improvement of Small-sized Palm Oil Industry. In 2000, the Chaipattana Foundation was commissioned by His Majesty the King to give financial support to the study on the use of palm oil. He also suggested that the use of palm oil in the diesel engine be experimented at some areas that people want to join the projects. The Chaipattana Foundation, in collaboration with ORDPB and Prince of Songkhla University, had shaped a guideline for the implementation of two following sub-projects in response to His Majesty's initiative. As the result, the Chaipattana Foundation in 2006 began to develop an area into the Vegetable Oil Refining and Bio-diesel Integrated Production Plant, which now serves as a model for environmentally-friendly bio-diesel production ("Energy Development", 2016).

Artificial rain

The project that is concerned about the suffering of the people in areas took place to solve the shortage of water for consumption and agriculture. Under the Chipattane

Foundation, the Ministry of Agriculture, and Air Force taking the initiative in the rain seriously, cabinet has agreed to include the operating rain with the Royal Flying agriculture into single units called "The rain and the flies Agriculture." And now, the rain that the King has bestowed on works since 1955 has been a tremendous benefit to the citizens. The root of technology about artificial rain causes directly to develop higher technology.

Production of information and technology society

Fritjof Capra begins by outlining and tracing the history of science and economics, highlighting flaws in the Cartesian, Newtonian, and reductionist paradigms which have come to light in the context of more contemporary empirical understanding of the physical sciences. He narrates how such viewpoints have grown inadequate to guide human behavior and policy with regard to modern technology and ecology, then argues that science needs to develop the concepts and insights of holism and systems theory to solve society's complex problems ("Fritjof Capra", 2016). The evidences supporting his conclusion can be found in the crucial subjects in present. In physics, quantum theory that explains the nature in subatomic level could not work well with Einstein's theory that explains the nature in celestial level. In economics, microeconomics that explains the economy phenomena of individual could not work well with macroeconomics that explains the economy phenomena of the whole society. Many physicists and economists are trying to indicate some weak points in their subjects; however, the warning was ignored by mainstream physicists and economists.

In addition, right now, the western countries are trying hard to control information sent to people around the world, some leading countries try to hide special information that they want people around the world to think about it. The information about UFOs or life or technology outer space at Air Technical Intelligence Center in Wright-Patterson Air Force Base in Ohio State of The United State of America, for example, was rejected by the government that viewed all evidences and information as the "things" that people in the countries should not know.

Much information was restricted to access. Many countries are trying to hack the information. Snowden and WikiLeaks are the two cases of trying of leading countries to control information of the world. The system of information was established by USA to judge what is right and what is wrong for people around the world. Moreover, the leading countries are holding the most advanced "root" of technology in their hands to protect themselves that are viewed as reasonable and to control the other countries that are viewed as reasonable as well.

The government that wants to develop a country to information and technology society is supposed to focus to the root of their own technology to support their people who develop their technology.

In Thailand, for instance, in the past, physicians ignored the ways of healing people by using natural herbs; many traditional physicians and their knowledge of healing were viewed as useless. In present, some physicians are still believe that their ways of healing people is the best, but some begins to think of the beneficial results if they know more about their present ways and their ancestors' ways. In China, acupuncture treatment in present was viewed as the alternative ways to healing people and was studied in many countries including Thailand.

The treatment is the Chinese's root of technology of healing that makes a large amount of money to China and Chinese's physicians around the world. Many western physicians rejected Chinese's acupuncture treatment because they did not have any ideas about the treatment. They also believed that they were the experts of healing treatment; however, when they found the treatment, they were in the status of finding paradigm shift. With the law of paradigm shift, when we are finding the situation of paradigm shift, we are be pushed to the starting point of the new paradigm. For example, when we were professors in physics or chemistry or and biology and were requested to explain artificial rain in Thailand, we will be pushed to the starting point of the technology. We are new students in the class of artificial rain, not experts in that new field.

If Chinese believed that western scientists were always the best in their subjects and ignored all of ancestor's knowledge, right now, they would have to pay the huge amount of money to import all of western medicine to use for Chinese. The trade deficit of medicine from aboard would be one of the causes of the poverty in China and the lack of root of development, China would become the one of countries that had modern styles of living without development.

The successful case study: His Majesty the King's Initiative on war ship construction

His Majesty the King granted an initiative for Royal Thai navy to construct Thai war ships. The explicit and implicit knowledge of Thai navy took place step by step after the first war ship had been constructed completely. After that, the bigger and bigger war ships have been constructed continuously for 30 years. The direct benefit such as decreasing Thailand's budget for buying new war ships and increasing both explicit and implicit knowledge is real root of technology that will be grown in Thai people like growing trees continuously. Both failure and achievement always give the stronger root of technology for society. And the society will leave the situation of modern without development.

Production of wisdom society: happy society

An obstacle that breaks the speed of development in countries is moral problems. Men are one of resources called human resources. Men may be viewed as "machines making money." In case of lack of moral in a society such as people' s commit suicide, people kill each other, people get lazy to do works, people are alcoholic, people cheat each other, the whole society will lose their opportunities to be happier. The government that wants people in a country to be happier needs to think of the happiness that occurs at the mind of people while their development of materialism is growing. In other words, development of people's mind has to be done at the same time of development of people's goods and services.

The society will get peace when these conditions have been done as follow: (1) people love and give mercy to each other, (2) people give the good to each other, (3) people' s honesty to their spouse, (4) people do not cheat each other, (5) people have their fulfill mind to do things. In contrast, the society will get more conflict when these conditions have been done as follow: when people hate and kill each other, the total values of society will decrease. When people hate and kill themselves, the whole values of society will also decrease.

Theory of consumer behavior

Many present economists believe that production with maximization of profit is so good. What is wrong if producers can produce mass products for buyers to buy cheap products? They believe also that the society will get "win-win strategy." The answer is in these questions "whether our children had right to use the resources of their country?" "If his generation used all natural resources without thinking of the people in the next generation, will children in the next generation complain what this generation did for them? "If our ancestors did not leave the valuable things for us to use, will our generation respect to their doing or not?" "Is the consumption with thinking of our children good for us to do?" Present economists believe that maximized utility is reasonable for us to think about when we try to consume somethings. Consumption economically is appropriate for us to do because we could not full ocean with a large amount of stones.

The wealth of nation

The objective of economic study is to search for the ways to get the wealth of the nation. The wealth of nation will be reached when people get peace, prosperous, and wellbeing ("COMMITTEE MEMBERS OF THE CHAIPATTANA FOUNDATION", 2016). To make these three economy destinations, these four conditions have to be done completely as follow: (1) people make money legally which their government can help them to “fish”, (2) people use their money economically for their present and future, (3) people keep money and resources with thinking of their future and their next generation, (4) people make good friends.

The first to third conditions were discussed in five steps of the theory of production and the theory of consumer’s behavior respectively. The last condition will be discussed in this section as follow. Right now, all of countries spend government expenditure for their armies. All of money is enough for the poor to have good opportunity to get jobs for making money. Unfortunately, their money is being used for winning the game of future war. The United State of America, NATO, USSR, and China believe that when the war begins they will win. In the past, World War I and II showed clearly that no one win in game of war including the winners. War did not make peace, wellbeing, and prosperous. Foreign policies of super power countries especially USA, NATO, USSR and China can make peace or war. If they try to make peace, they will get new friends, but if they try to make war, they will get new enemy. If these countries make friends to others, they will decrease the government budget on military and increase the budget on people’s working, four living factors.

The application of the new economic theory

Food stamp in Los Angeles

The objective of food stamp project in Los Angeles is to help people especially young moms who feed their kids alone. The project will not be burden of society if the governor helps the mothers to get jobs to earn themselves and their kids.

Homeless in Los Angeles

If the governor of Los Angeles has jobs got for the homeless people to earn themselves, he will relax the burden of the state and increase happiness to all people.

Unemployment in Los Angeles

If the governor has new public jobs created such as making new roads to increase jobs to help people to earn themselves or train new jobs that people want to do for earning their life, he will relax the burden of the state and increase happiness to all people as well.

Trump's policy about sending illegal foreigners back home

If President Trump and the governors of states have new public jobs created such as making new roads to increase jobs to help people to earn themselves or train new jobs that people want to do for earning their life, he will relax the burden of the state and increase happiness to all people as well.

Trump's policy about USSR

If two presidents of two countries make friends, the countries can reduce the government expenditure on military and use the money for people to get jobs to earn themselves. They will decrease the burden of the states on military and increase happiness to all people as well.

The conclusion of the chapter

Everything depends on time. The second characteristic of time is that values of time can be calculated and indicated the exact positions matching in all numbers on number lines by the time equation. The chart of crucial contents of this book can be written as the following.

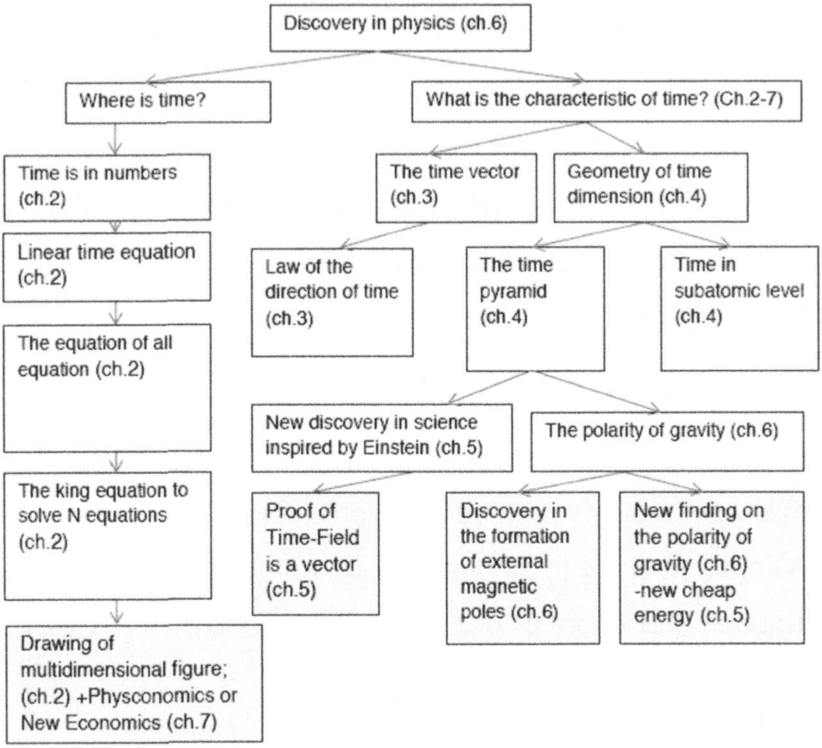

Figure 93: The chart of the results of my research with applied mathematics.

The new price model with time can explain the real world prices in the real world. The Exchanging Line was created to explain the prices in the real world as well. Wage, Rent, interest, and profit are prices with different names. The price in the real world is data, not money. In price system, when exchanging process between buyers and sellers is taking place, only money and products or services are transferred to opposite sides. There is no price at that moment we can see because of price as data. The model can link microeconomics and macroeconomics together as well.

The model of exchanging line with time develops economic theories to scientific paradigm. The exogenous variables are not the limitation of any economic models in the future study because the exogenous variables become the endogenous variables of the new mathematical price model with positive economics and inductive method. It can be used for explaining the world of neoclassical theories with normative economics and deductive method as well.

Keynes' theory, which could not explain how high unemployment and high inflation occurs simultaneously, was created by Keynes using normative economics. Keynes believed that the equilibrium of interest rate was made from money supply and demand for money in economy system. With time as the linkage between micro- and macroeconomics, price is the real linkage between money and product markets with mathematical evidence, $P_t = M_{Tt}/Q_{Tt}$. The interest rates depend on the owners of many commercial banks, which are running business on money. There is great business competition between the various banks fighting for first place; this makes bankers choose the different ways on changing interest rates on loan and saving. In the real world economy, the interest rates of commercial banks do not depend on money supply and demand for money in economy system.

If government increases money supply in economy system without increasing products and services; prices of products will increase. If government cannot control low inflation rate; many economy problems will occur such as increasing unemployment. These are the real causes that make high unemployment and high inflation occurring at the same time.

Owners of firms, which normally have higher bargaining power than their employees have, will buy the time of their employees for working for organization. In many cases, employees often have higher bargaining power than the employers have.

If government wants to increase money supply in economy system, one of the best ways, which government should do, is that government has to increase products or services of many poor people in many ways. As the result, many poor people will increase higher consumption of products that will make producers produce their products, and wholesalers and retailers want to buy their products for selling to buyers.

If government uses budget deficit without increasing products or services of many people; budget deficit makes higher debts that will be the burden of the society in the long run for paying back the debts, capital and interest rate.

There are many strategies to increase money in economy system because of new price function giving many guidelines to do. Time is still the crucial dimension that can help economists to develop future economic theory for explaining the real world economy.

The impact of this research affects many fields directly and indirectly. If we overlook the time direction, many serious problems will occur. In the other hand, a big change in fundamental concept of physics including mathematics has been coming.

The duty of economists is pointing of the right direction for policy-makers to control economy and to reach the wealth of nation. In addition, the new theory in economics can be created for answers how to reach the wealth of nation every day. The new theory of H.M. the King Rama IX that was the king of Thailand is the good example for economists to explain why the projects can increase average Thai income per capita and how to get the victory of development in the higher level step by step.

Finally, the case study of projects initiated by His Majestic the King Rama IX of Thailand was explained by using new price model. In my opinion, the objective of king's projects is helping people to "fish" with two keys words, four categories of development, and 10 main kinds of work. The model has showed the linkage between micro-and macroeconomics that present neoclassical economists have never done. In addition, the model can demonstrate a new strategy to increase economy growth in the real world continuously. It also changes economics from social science to scientific subjects.

I really hope that in the near future economists can give valuable directions for people around the world to reach their wealth of nations together without leaving any countries behind. At that point, the happiness of people around the world will be the gifts for economists who are the scientists of societies that make people get richer and happier. The victory of development is peace, prosperity, and national wellbeing ("COMMITTEE MEMBERS OF THE CHAIPATTANA FOUNDATION", 2016).

Mathematical Proof

Chapter 2

1. The proof of S = f (t)

From the equation of distance:

$$S = v\,t$$

$$\frac{dS}{dt} = \frac{d(vt)}{dt}$$

Given v = constant

$$\frac{dS}{dt} = \frac{vd(t)}{dt} + \frac{td(v)}{dt}$$

$$= V(1) + t(0)$$

$$= V$$

$$S = \left(\frac{dS}{dt}\right) t$$

$$= 0 + \left(\frac{dS}{dt}\right) t$$

The relation between S and t is linear, or S^1 and t^1; as the result,

$$\frac{dS}{dt} = \left(\frac{\Delta S}{\Delta t}\right)$$

$$S = 0 + \left(\frac{\Delta S}{\Delta t}\right) t$$

$$S = f\,(t) \quad \text{...Q.E.D}$$

2. We can calculate the value of time matching with the fifth determined number (500) as follow:

The first method by using S and v;

$$t_{at500} = \frac{S}{v} = \frac{100 \text{ inches}}{2 \text{ inches per second}} = \frac{100 \text{ inches}}{\frac{2 \text{ inches}}{1 \text{ second}}} = 50 \text{ seconds}$$

The second method by using N, R, and v;

$$R = \frac{|N \text{ tons}|}{S \text{ inches}} = \frac{|500 \text{ tons}|}{100 \text{ inches}} = \frac{500 \text{ tons}}{100 \text{ inches}} = \frac{5 \text{ tons}}{1 \text{ inch}}$$

$$t_{at500} = \frac{|N|}{R.v} = \frac{|500 \text{ tons}|}{(5 \text{ tons per inch})(2 \text{ inches per second})} = \frac{500 \text{ tons}}{\left(\frac{5 \text{ tons}}{1 \text{ inch}}\right)\left(\frac{2 \text{ inches}}{1 \text{ second}}\right)} = 50 \text{ seconds}$$

3. We can calculate the value of time matching with the fifth determined number (-500) as follow:

The first method by using S and v;

$$t_{at-500} = \frac{S}{v} = \frac{100 \text{ inches}}{2 \text{ inches per second}} = \frac{100 \text{ inches}}{\frac{2 \text{ inches}}{1 \text{ second}}} = 50 \text{ seconds}$$

The second method by using N, R, and v;

$$R = \frac{|N|}{S \text{ inches}} = \frac{|-500|}{100 \text{ inches}} = \frac{500}{100 \text{ inches}} = \frac{5}{1 \text{ inch}}$$

$$t_{at-500} = \frac{|N|}{R.v} = \frac{|-500|}{(5 \text{ per inch})(2 \text{ inches per second})} = \frac{500}{\left(\frac{5}{1 \text{ inch}}\right)\left(\frac{2 \text{ inches}}{1 \text{ second}}\right)} = 50 \text{ seconds}$$

Chapter 3

1. We can calculate the time value on the y-axis.

The first method by using S and v;

$$t_y = \frac{S_y}{v_y} = \frac{1 \text{ inch}}{1 \text{ inch per second}} = \frac{1 \text{ inch}}{\frac{1 \text{ inch}}{1 \text{ second}}} = 1 \text{ second}$$

The second method by using N, R, and v;

$$R_y = \frac{|N_y|}{S_y \text{ inches}} = \frac{1}{1 \text{ inch}} = \frac{1}{1 \text{ inch}}$$

$$t_y = \frac{|N_y|}{R_y . v_y} = \frac{1}{(1 \text{ per inch})(1 \text{ inch per second})} = \frac{1}{\left(\frac{1}{1 \text{ inch}}\right)\left(\frac{1 \text{ inch}}{1 \text{ second}}\right)} = 1 \text{ second}$$

2. We can calculate the time value on the z-axis.

The first method by using S and v;

$$t_z = \frac{S_z}{v_z} = \frac{1 \text{ inch}}{1 \text{ inch per second}} = \frac{1 \text{ inch}}{\frac{1 \text{ inch}}{1 \text{ second}}} = 1 \text{ second}$$

The second method by using N, R, and v;

$$R_z = \frac{|N_z|}{S_z \text{ inches}} = \frac{1}{1 \text{ inch}} = \frac{1}{1 \text{ inch}}$$

$$t_z = \frac{|N_z|}{R_z . v_z} = \frac{1}{(1 \text{ per inch})(1 \text{ inch per second})} = \frac{1}{\left(\frac{1}{1 \text{ inch}}\right)\left(\frac{1 \text{ inch}}{1 \text{ second}}\right)} = 1 \text{ second}$$

Table 8: The resultant time directions in the eight parts of the three-dimensional body with the well-known numbers of vectors

No. of the part	The resultant time directions explained by using symbols of vector (Boonsri, 2002)	The magnitudes of the resultant time directions explained by using symbols of vector (Boonsri, 2002)
1	$\mathbf{A}_{oc1} = A_1\mathbf{i} + A_2\mathbf{j} + A_3\mathbf{k}$	$A_{oc1} = \sqrt{(A_1)^2 + (A_2)^2 + (A_3)^2}$
2	$\mathbf{A}_{oc2} = A_1\mathbf{i} + A_2\mathbf{j} - A_3\mathbf{k}$	$A_{oc2} = \sqrt{(A_1)^2 + (A_2)^2 + (-A_3)^2}$
3	$\mathbf{A}_{oc3} = -A_1\mathbf{i} + A_2\mathbf{j} - A_3\mathbf{k}$	$A_{oc3} = \sqrt{(-A_1)^2 + (A_2)^2 + (-A_3)^2}$
4	$\mathbf{A}_{oc4} = -A_1\mathbf{i} + A_2\mathbf{j} + A_3\mathbf{k}$	$A_{oc4} = \sqrt{(-A_1)^2 + (A_2)^2 + (A_3)^2}$
5	$\mathbf{A}_{oc5} = A_1\mathbf{i} - A_2\mathbf{j} + A_3\mathbf{k}$	$A_{oc5} = \sqrt{(A_1)^2 + (-A_2)^2 + (A_3)^2}$
6	$\mathbf{A}_{oc6} = A_1\mathbf{i} - A_2\mathbf{j} - A_3\mathbf{k}$	$A_{oc6} = \sqrt{(A_1)^2 + (-A_2)^2 + (-A_3)^2}$

7	$\mathbf{A}_{oc7}$ = -A_1**i** - A_2**j** - A_3**k**	$A_{oc7} = \sqrt{(-A_1)^2 + (-A_2)^2 + (-A_3)^2}$
8	$\mathbf{A}_{oc8}$ = -A_1**i** - A_2**j** + A_3**k**	$A_{oc8} = \sqrt{(-A_1)^2 + (-A_2)^2 + (A_3)^2}$

Table 9: The resultant time directions in the eight parts of the three-dimensional body with the determined numbers in this book that match with the numbers in Table 8.

No. of the part	The resultant time directions explained by using the determined numbers used in this book	The magnitudes of the resultant time directions explained by using the determined numbers in this book (Boonsri, 2016)
1	$\mathbf{N}_{oc1}$ = N_x**i** + N_y**j** + N_z**k**	$N_{oc1} = \sqrt{(N_x)^2 + (N_y)^2 + (N_z)^2}$
2	$\mathbf{N}_{oc2}$ = N_x**i** + N_y**j** – N_z**k**	$N_{oc2} = \sqrt{(N_x)^2 + (N_y)^2 + (-N_z)^2}$
3	$\mathbf{N}_{oc3}$ = –N_x**i** + N_y**j** – N_z**k**	$N_{oc3} = \sqrt{(-N_x)^2 + (N_y)^2 + (-N_y)^2}$
4	$\mathbf{N}_{oc4}$ = –N_x**i** + N_y**j** + N_z**k**	$N_{oc4} = \sqrt{(-N_x)^2 + (N_y)^2 + (N_z)^2}$
5	$\mathbf{N}_{oc5}$ = N_x**i** – N_y**j** + N_z**k**	$N_{oc5} = \sqrt{(N_x)^2 + (-N_y)^2 + (N_z)^2}$
6	$\mathbf{N}_{oc6}$ = N_x**i** – N_y**j** – N_z**k**	$N_{oc6} = \sqrt{(N_x)^2 + (-N_y)^2 + (-N_z)^2}$
7	$\mathbf{N}_{oc7}$ = –N_x**i** – N_y**j** – N_z**k**	$N_{oc7} = \sqrt{(-N_x)^2 + (-N_y)^2 + (-N_z)^2}$
8	$\mathbf{N}_{oc8}$ = –N_x**i** – N_y**j** + N_z**k**	$N_{oc8} = \sqrt{(-N_x)^2 + (-N_y)^2 + (N_z)^2}$

Table 10: The resultant time directions in the eight parts of the three-dimensional body in the form showing distances that match with the numbers in Table 8

No. of a part	The resultant time directions explained by using symbols showing distances	The magnitudes of the resultant time directions explained by using symbols showing distances
1	$\mathbf{S}_{oc1} = (S_xR_x)\mathbf{i} + (S_yR_y)\mathbf{j} + (S_zR_z)\mathbf{k}$	$S_{oc1} = \sqrt{(S_xR_x)^2 + (S_yR_y)^2 + (S_zR_z)^2}$
2	$\mathbf{S}_{oc2} = (S_xR_x)\mathbf{i} + (S_yR_y)\mathbf{j} - (S_zR_z)\mathbf{k}$	$S_{oc2} = \sqrt{(S_xR_x)^2 + (S_yR_y)^2 + (-S_zR_z)^2}$
3	$\mathbf{S}_{oc3} = -(S_xR_x)\mathbf{i} + (S_yR_y)\mathbf{j} - (S_zR_z)\mathbf{k}$	$S_{oc3} = \sqrt{(-S_xR_x)^2 + (S_yR_y)^2 + (-S_zR_z)^2}$
4	$\mathbf{S}_{oc4} = -(S_xR_x)\mathbf{i} + (S_yR_y)\mathbf{j} + (S_zR_z)\mathbf{k}$	$S_{oc4} = \sqrt{(-S_xR_x)^2 + (S_yR_y)^2 + (S_zR_z)^2}$
5	$\mathbf{S}_{oc5} = (S_xR_x)\mathbf{i} - (S_yR_y)\mathbf{j} + (S_zR_z)\mathbf{k}$	$S_{oc5} = \sqrt{(S_xR_x)^2 + (-S_yR_y)^2 + (S_zR_z)^2}$
6	$\mathbf{S}_{oc6} = (S_xR_x)\mathbf{i} - (S_yR_y)\mathbf{j} - (S_zR_z)\mathbf{k}$	$S_{oc6} = \sqrt{(S_xR_x)^2 + (-S_yR_y)^2 + (-S_zR_z)^2}$
7	$\mathbf{S}_{oc7} = -(S_xR_x)\mathbf{i} - (S_yR_y)\mathbf{j} - (S_zR_z)\mathbf{k}$	$S_{oc7} = \sqrt{(-S_xR_x)^2 + (-S_yR_y)^2 + (-S_zR_z)^2}$
8	$\mathbf{S}_{oc8} = -(S_xR_x)\mathbf{i} - (S_yR_y)\mathbf{j} + (S_zR_z)\mathbf{k}$	$S_{oc8} = \sqrt{(-S_xR_x)^2 + (-S_yR_y)^2 + (S_zR_z)^2}$

Table 11: The resultant time directions in the eight parts of the three-dimensional body in the form showing the values of time that match with the numbers in Table 8

No. of each part	The resultant time directions explained by using symbols showing values of time	The magnitudes of the resultant time directions explained by using symbols showing values of time (Boonsri, 2016)
1	$\mathbf{t}_{oc1} = (v_x t_x R_x)\mathbf{i} + (v_y t_y R_y)\mathbf{j} + (v_z t_z R_z)\mathbf{k}$	$t_{oc1} = \sqrt{(v_x t_x R_x)^2 + (v_y t_y R_y)^2 + (v_z t_z R_z)^2}$
2	$\mathbf{t}_{oc2} = (v_x t_x R_x)\mathbf{i} + (v_y t_y R_y)\mathbf{j} - (v_z t_z R_z)\mathbf{k}$	$t_{oc2} = \sqrt{(v_x t_x R_x)^2 + (v_y t_y R_y)^2 + (-v_z t_z R_z)^2}$
3	$\mathbf{t}_{oc3} = -(v_x t_x R_x)\mathbf{i} + (v_y t_y R_y)\mathbf{j} - (v_z t_z R_z)\mathbf{k}$	$t_{oc3} = \sqrt{(-v_x t_x R_x)^2 + (v_y t_y R_y)^2 + (-v_z t_z R_z)^2}$
4	$\mathbf{t}_{oc4} = -(v_x t_x R_x)\mathbf{i} + (v_y t_y R_y)\mathbf{j} + (v_z t_z R_z)\mathbf{k}$	$t_{oc4} = \sqrt{(-v_x t_x R_x)^2 + (v_y t_y R_y)^2 + (v_z t_z R_z)^2}$
5	$\mathbf{t}_{oc5} = (v_x t_x R_x)\mathbf{i} - (v_y t_y R_y)\mathbf{j} + (v_z t_z R_z)\mathbf{k}$	$t_{oc5} = \sqrt{(v_x t_x R_x)^2 + (-v_y t_y R_y)^2 + (v_z t_z R_z)^2}$
6	$\mathbf{t}_{oc6} = (v_x t_x R_x)\mathbf{i} - (v_y t_y R_y)\mathbf{j} - (v_z t_z R_z)\mathbf{k}$	$t_{oc6} = \sqrt{(v_x t_x R_x)^2 + (-v_y t_y R_y)^2 + (-v_z t_z R_z)^2}$

7	$\mathbf{t}_{oc7} = -(v_x t_x R_x)\mathbf{i} - (v_y t_y R_y)\mathbf{j} - (v_z t_z R_z)\mathbf{k}$	$t_{oc7} = \sqrt{(-v_x t_x R_x)^2 + (-v_y t_y R_y)^2 + (-v_z t_z R_z)^2}$
8	$\mathbf{t}_{oc8} = -(v_x t_x R_x)\mathbf{i} - (v_y t_y R_y)\mathbf{j} + (v_z t_z R_z)\mathbf{k}$	$t_{oc8} = \sqrt{(-v_x t_x R_x)^2 + (-v_y t_y R_y)^2 + (v_z t_z R_z)^2}$

Chapter 4

The resultant directions of the six time-fields were showed. The data were used to solve as follow:

$\mathbf{A}_{oc1} = A_1\mathbf{i} + A_2\mathbf{j} + A_3\mathbf{k}$, $\mathbf{A}_{oc2} = A_1\mathbf{i} + A_2\mathbf{j} - A_3\mathbf{k}$,

$\mathbf{A}_{oc3} = -A_1\mathbf{i} + A_2\mathbf{j} - A_3\mathbf{k}$, $\mathbf{A}_{oc2} = -A_1\mathbf{i} + A_2\mathbf{j} + A_3\mathbf{k}$,

$\mathbf{A}_{oc5} = A_1\mathbf{i} - A_2\mathbf{j} + A_3\mathbf{k}$, $\mathbf{A}_{oc6} = A_1\mathbf{i} - A_2\mathbf{j} - A_3\mathbf{k}$

$\mathbf{A}_{oc7} = -A_1\mathbf{i} - A_2\mathbf{j} - A_3\mathbf{k}$, $\mathbf{A}_{oc6} = -A_1\mathbf{i} - A_2\mathbf{j} + A_3\mathbf{k}$

The resultant direction of the right, left, upper, bottom, front, and back time-fields in the form of numbers respectively:

1. The resultant direction of the right time-field in the form of numbers:

$\mathbf{A}_{PAright}$ = the resultant direction of the right time-field

= the resultant vector of the four resultant directions of time

$= \mathbf{A}_{oc1} + \mathbf{A}_{oc2} + \mathbf{A}_{oc5} + \mathbf{A}_{oc6}$

$= (A_1+A_1+A_1+A_1)\,\mathbf{i} + (A_2+A_2-A_2-A_2)\,\mathbf{j} + (A_3-A_3+A_3-A_3)\,\mathbf{k}$

$= 4A_1\mathbf{i} + 0\,\mathbf{j} + 0\,\mathbf{k} = 4A_1\mathbf{i}$

= direction along the principal x-axis with positive side

= $4(v_x t_x R_x)$ **i**

= direction along the principal x-axis with positive side in the form showing time

$A_{PAright}$ = the magnitude of the resultant direction of the right time-field

= the magnitude of the resultant vector of the right time-field

= $|\mathbf{A}_{oc1} + \mathbf{A}_{oc2} + \mathbf{A}_{oc5} + \mathbf{A}_{oc6}|$

= $\sqrt{(4A_1)^2 + (0)^2 + (0)^2}$

= $\sqrt{16A_1^2}$

= $4A_1$

= the same magnitude as $A_{PAright}$

= $4(v_x t_x R_x)$ = the same magnitude as A_{PAleft} in the form showing time

2. The resultant direction of the left time-field in the form of numbers:

$\mathbf{A}_{PAleft}$ = the resultant direction of the left time-field

= the resultant vector of the four resultant directions of time

= $\mathbf{A}_{oc3} + \mathbf{A}_{oc4} + \mathbf{A}_{oc7} + \mathbf{A}_{oc8}$

= $(-A_1-A_1-A_1-A_1)\,\mathbf{i} + (A_2+A_2-A_2-A_2)\,\mathbf{j} + (-A_3+A_3-A_3+A_3)\,\mathbf{k}$

= $-4A_1\mathbf{i} + 0\,\mathbf{j} + 0\,\mathbf{k} = -4A_1\mathbf{i}$

= direction along the principal x-axis with negative side

= $-4(v_x t_x R_x)$ **i**

= direction along the principal x-axis with negative side in the form showing time

A_{PAleft} = the magnitude of the resultant direction of the left time-field

= the magnitude of the resultant vector of the left time-field

= $|\mathbf{A_{oc3}} + \mathbf{A_{oc4}} + \mathbf{A_{oc7}} + \mathbf{A_{oc8}}|$

= $\sqrt{(-4A_1)^2 + (0)^2 + (0)^2}$

= $\sqrt{16A_1^2}$

= $4A_1$

= the same magnitude as $A_{PAright}$

= $4(v_x t_x R_x)$ = the same magnitude as $A_{PAright}$ in the form showing time

3. The resultant direction of the upper time-field in the form of numbers:

$\mathbf{A_{PAupper}}$ = the resultant direction of the upper time-field

= the resultant vector of the four resultant directions of time

= $\mathbf{A_{oc1}} + \mathbf{A_{oc2}} + \mathbf{A_{oc3}} + \mathbf{A_{oc4}}$

= $(A_1+A_1-A_1-A_1)\,\mathbf{i} + (A_2+A_2+A_2+A_2)\,\mathbf{j} + (A_3-A_3-A_3+A_3)\,\mathbf{k}$

= $0\,\mathbf{i} + 4A_2\,\mathbf{j} + 0\,\mathbf{k} = 4A_2\,\mathbf{j}$

= direction along the principal y-axis with positive side

	=	$4(v_x t_x R_x)$ **j**
	=	direction along the principal y-axis with positive side in the form showing time
$A_{PAupper}$	=	the magnitude of the resultant direction of the upper time-field
	=	the magnitude of the resultant vector of the upper time-field
	=	$\lvert \mathbf{A}_{oc1} + \mathbf{A}_{oc2} + \mathbf{A}_{oc3} + \mathbf{A}_{oc4} \rvert$
	=	$\sqrt{(0)^2 + (4A_2)^2 + (0)^2}$
	=	$\sqrt{16A_2^2}$
	=	$4A_2$
	=	the same magnitude as $A_{PAbottom}$
	=	$4(v_y t_y R_y)$ = the same magnitude as $A_{PAbottom}$ in the form showing time

4. The resultant direction of the bottom time-field in the form of numbers:

$\mathbf{A}_{PAbottom}$	=	the resultant direction of the bottom time-field
	=	the resultant vector of the four resultant directions of time
	=	$\mathbf{A}_{oc5} + \mathbf{A}_{oc6} + \mathbf{A}_{oc7} + \mathbf{A}_{oc8}$
	=	$(A_1+A_1-A_1-A_1)\,\mathbf{i} + (-A_2-A_2-A_2-A_2)\,\mathbf{j} + (A_3-A_3-A_3+A_3)\,\mathbf{k}$
	=	$0\,\mathbf{i} - 4A_2\,\mathbf{j} + 0\,\mathbf{k} = -4A_2\,\mathbf{j}$
	=	direction along the principal y-axis with negative side

= $-4(v_x t_x R_x)$ **j**

= direction along the principal y-axis with negative side in the form showing time

$A_{PAbottom}$ = the magnitude of the resultant direction of the bottom time-field

= the magnitude of the resultant vector of the bottom time-field

= $|\mathbf{A_{oc5}} + \mathbf{A_{oc6}} + \mathbf{A_{oc7}} + \mathbf{A_{oc8}}|$

= $\sqrt{(0)^2 + (-4A_2)^2 + (0)^2}$

= $\sqrt{16A_2^2}$

= $4A_2$

= the same magnitude as $A_{PAupper}$

= $4(v_y t_y R_y)$ = the same magnitude as $A_{PAupper}$ in the form showing time

5. The resultant direction of the front time-field in the form of numbers:

$\mathbf{A_{PAfront}}$ = the resultant direction of the front time-field

= the resultant vector of the four resultant directions of time

= $\mathbf{A_{oc1}} + \mathbf{A_{oc4}} + \mathbf{A_{oc5}} + \mathbf{A_{oc8}}$

= $(A_1-A_1+A_1-A_1)$ **i** + $(A_2+A_2-A_2-A_2)$ **j**+ $(A_3+A_3+A_3+A_3)$ **k**

= 0 **i** + 0 **j** + $4A_1$ **k** = $4A_3$ **k**

= direction along the principal z-axis with positive side

	=	$4(v_z t_z R_z)$ **k**
	=	direction along the principal z-axis with positive side in the form showing time
$A_{PAfront}$	=	the magnitude of the resultant direction of the front time-field
	=	the magnitude of the resultant vector of the front time-field
	=	$\lvert \mathbf{A}_{oc1} + \mathbf{A}_{oc4} + \mathbf{A}_{oc5} + \mathbf{A}_{oc8} \rvert$
	=	$\sqrt{(0)^2 + (0)^2 + (4A_3)^2}$
	=	$\sqrt{16A_3^2}$
	=	$4A_3$
	=	the same magnitude as A_{PAback}
	=	$4(v_z t_z R_z)$ = the same magnitude as A_{PAback} in the form showing time

6. The resultant direction of the back time-field in the form of numbers:

$\mathbf{A}_{PAback}$	=	the resultant direction of the back time-field
	=	the resultant vector of the four resultant directions of time
	=	$\mathbf{A}_{oc2} + \mathbf{A}_{oc3} + \mathbf{A}_{oc6} + \mathbf{A}_{oc7}$
	=	$(A_1 - A_1 + A_1 - A_1)$ **i** + $(A_2 + A_2 - A_2 - A_2)$ **j** + $(-A_3 - A_3 - A_3 - A_3)$ **k**
	=	0 **i** + 0 **j** - $4A_1$ **k** = $-4A_3$ **k**
	=	direction along the principal z-axis with negative side

= $-4(v_z t_z R_z)$ **k**

= direction along the principal z-axis with negative side in the form showing time

A_{PAback} = the magnitude of the resultant direction of the back time-field

= the magnitude of the resultant vector of the back time-field

= $|\mathbf{A_{oc2}} + \mathbf{A_{oc3}} + \mathbf{A_{oc6}} + \mathbf{A_{oc7}}|$

= $\sqrt{(0)^2 + (0)^2 + (-4A_3)^2}$

= $\sqrt{16A_3^2}$

= $4A_3$

= the same magnitude as $A_{PAfront}$

= $4(v_z t_z R_z)$ = the same magnitude as $A_{PAfront}$ in the form showing time

7. To find the angle 109.4712°, for example, between $\mathbf{A_{oc1}} = A_1\mathbf{i} + A_2\mathbf{j} + A_3\mathbf{k}$ and $\mathbf{A_{oc6}} = A_1\mathbf{i} - A_2\mathbf{j} - A_3\mathbf{k}$, it can be solved as follow:

$\mathbf{A_{oc1}}.\mathbf{A_{oc6}}$ = $(A_1)(A_1)+(A_2)(-A_2)+(A_3)(-A_3)$ = $(A_1)^2 - (A_2)^2 - (A_3)^2$

These two vectors with the bold symbols, $\mathbf{A_{oc1}}.\mathbf{A_{oc6}}$, are called dot product or scalar product.

But in all cases in this chapter, $A_1 = A_2 = A_3$; as a result,

$\mathbf{A_{oc1}}.\mathbf{A_{oc6}}$ = $(A_1)(A_1)+(A_2)(-A_2)+(A_3)(-A_3)$ = $(A_1)^2 - (A_2)^2 - (A_3)^2$

$$= (A_1)^2 - (A_1)^2 - (A_1)^2 = -(A_1)^2 \text{ or}$$

$$= (A_2)^2 - (A_2)^2 - (A_2)^2 = -(A_2)^2 \text{ or}$$

$$= (A_3)^2 - (A_3)^2 - (A_3)^2 = -(A_3)^2$$

And these two magnitudes that are shown as symbols without boldness, $A_{oc1}.A_{oc6}$, are solved as follow:

$$A_{oc1} = \sqrt{(A_1)^2 + (A_2)^2 + (A_3)^2}$$

$$= \sqrt{(A_1)^2 + (A_1)^2 + (A_1)^2} = \sqrt{3}\, A_1$$

$$= \sqrt{(A_2)^2 + (A_2)^2 + (A_2)^2} = \sqrt{3}\, A_2$$

$$= \sqrt{(A_3)^2 + (A_3)^2 + (A_3)^2} = \sqrt{3}\, A_3$$

$$A_{oc6} = \sqrt{(A_1)^2 + (-A_2)^2 + (-A_3)^2}$$

$$= \sqrt{(A_1)^2 + (A_1)^2 + (A_1)^2} = \sqrt{3}\, A_1$$

$$= \sqrt{(A_2)^2 + (A_2)^2 + (A_2)^2} = \sqrt{3}\, A_2$$

$$= \sqrt{(A_3)^2 + (A_3)^2 + (A_3)^2} = \sqrt{3}\, A_3$$

$$\cos \omega_{16} = \frac{\mathbf{A_{oc1}}.\mathbf{A_{oc6}}}{A_{oc1}.\,A_{oc6}} = \frac{-(A_1)^2}{\sqrt{3}\ A_1 \,.\, \sqrt{3}\, A_1}$$

$$= \frac{-(v_x t_x R_x)^2}{\sqrt{3}(v_x t_x R_x).\sqrt{3}(v_x t_x R_x)} = \frac{-1}{3} = -0.3333$$

Hence

$$\omega_{16} = \cos^{-1}\left(\frac{-(A_1)^2}{\sqrt{3}\,A_1 \,.\, \sqrt{3}A_1}\right) = \cos^{-1}\left(\frac{-(A_2)^2}{\sqrt{3}\,A_2 \,.\, \sqrt{3}A_2}\right)$$

$$= \cos^{-1}\left(\frac{-(A_3)^2}{\sqrt{3}\,A_3 \,.\, \sqrt{3}A_3}\right) = \cos^{-1}\left(\frac{-1}{3}\right)$$

$$= \cos^{-1}(-0.333) = 109.4712°$$

With the same method, these cosine angles in the six time-fields can be solved as follow:

$$\cos \omega_{25} = \frac{\mathbf{A_{oc2}} \cdot \mathbf{A_{oc5}}}{A_{oc2}.\, A_{oc5}} = \frac{-(A_1)^2}{\sqrt{3}\, A_1 \,.\, \sqrt{3}\, A_1} = \frac{-1}{3}$$

$$\cos \omega_{38} = \frac{\mathbf{A_{oc3}} \cdot \mathbf{A_{oc8}}}{A_{oc3}.\, A_{oc8}} = \frac{-(A_1)^2}{\sqrt{3}\, A_1 \,.\, \sqrt{3}\, A_1} = \frac{-1}{3}$$

$$\cos \omega_{47} = \frac{\mathbf{A_{oc4}} \cdot \mathbf{A_{oc7}}}{A_{oc4}.\, A_{oc7}} = \frac{-(A_1)^2}{\sqrt{3}\, A_1 \,.\, \sqrt{3}\, A_1} = \frac{-1}{3}$$

$$\cos \omega_{13} = \frac{\mathbf{A_{oc1}} \cdot \mathbf{A_{oc3}}}{A_{oc1}.\, A_{oc3}} = \frac{-(A_1)^2}{\sqrt{3}\, A_1 \,.\, \sqrt{3}\, A_1} = \frac{-1}{3}$$

$$\cos \omega_{24} = \frac{\mathbf{A_{oc2}} \cdot \mathbf{A_{oc4}}}{A_{oc2}.\, A_{oc4}} = \frac{-(A_1)^2}{\sqrt{3}\, A_1 \,.\, \sqrt{3}\, A_1} = \frac{-1}{3}$$

$$\cos \omega_{57} = \frac{\mathbf{A_{oc5}} \cdot \mathbf{A_{oc7}}}{A_{oc5}.\, A_{oc7}} = \frac{-(A_1)^2}{\sqrt{3}\, A_1 \,.\, \sqrt{3}\, A_1} = \frac{-1}{3}$$

$$\cos \omega_{68} = \frac{\mathbf{A_{oc6}} \cdot \mathbf{A_{oc8}}}{A_{oc6}.\, A_{oc8}} = \frac{-(A_1)^2}{\sqrt{3}\, A_1 \,.\, \sqrt{3}\, A_1} = \frac{-1}{3}$$

$$\cos \omega_{18} = \frac{\mathbf{A_{oc1}} \cdot \mathbf{A_{oc8}}}{A_{oc1}.\, A_{oc8}} = \frac{-(A_1)^2}{\sqrt{3}\, A_1 \,.\, \sqrt{3}\, A_1} = \frac{-1}{3}$$

$$\cos \omega_{45} = \frac{\mathbf{A_{oc4}} \cdot \mathbf{A_{oc5}}}{A_{oc4}.\, A_{oc5}} = \frac{-(A_1)^2}{\sqrt{3}\, A_1 \,.\, \sqrt{3}\, A_1} = \frac{-1}{3}$$

$$\cos \omega_{27} = \frac{\mathbf{A_{oc2}} \cdot \mathbf{A_{oc7}}}{A_{oc2}.\, A_{oc7}} = \frac{-(A_1)^2}{\sqrt{3}\, A_1 \,.\, \sqrt{3}\, A_1} = \frac{-1}{3}$$

$$\cos \omega_{36} = \frac{\mathbf{A_{oc3}} \cdot \mathbf{A_{oc6}}}{A_{oc3}.\, A_{oc6}} = \frac{-(A_1)^2}{\sqrt{3}\, A_1 \,.\, \sqrt{3}\, A_1} = \frac{-1}{3}$$

Conclusion:

$$\cos 109.4712^{\circ} = \cos \omega_{13} = \cos \omega_{16} = \cos \omega_{18}$$

=	$\cos\omega_{24}$	=	$\cos\omega_{25}$	=	$\cos\omega_{27}$
=	$\cos\omega_{36}$	=	$\cos\omega_{38}$	=	$\cos\omega_{45}$
=	$\cos\omega_{47}$	=	$\cos\omega_{57}$	=	$\cos\omega_{68}$

8. To find the angle 70.5287°, for example, between $\mathbf{A}_{oc1} = A_1\mathbf{i} + A_2\mathbf{j} + A_3\mathbf{k}$ and $\mathbf{A}_{oc2} = A_1\mathbf{i} + A_2\mathbf{j} - A_3\mathbf{k}$, it can be solved as follow:

$$\mathbf{A}_{oc1}.\mathbf{A}_{oc2} = (A_1)(A_1)+(A_2)(A_2)+(A_3)(-A_3) = (A_1)^2 + (A_2)^2 - (A_3)^2$$

But in all cases in this chapter, $A_1 = A_2 = A_3$; as a result,

$$\mathbf{A}_{oc1}.\mathbf{A}_{oc2} = (A_1)(A_1)+(A_2)(A_2)+(A_3)(-A_3) = (A_1)^2 + (A_2)^2 - (A_3)^2$$

$$= (A_1)^2 + (A_1)^2 - (A_1)^2 = (A_1)^2 \quad \text{or}$$

$$= (A_2)^2 + (A_2)^2 - (A_2)^2 = (A_2)^2 \quad \text{or}$$

$$= (A_3)^2 + (A_3)^2 - (A_3)^2 = (A_3)^2$$

And these two magnitudes that are shown as symbols without boldness, $A_{oc1}.A_{oc2}$, are solved as follow:

$$A_{oc1} = \sqrt{(A_1)^2 + (A_2)^2 + (A_3)^2}$$

$$= \sqrt{(A_1)^2 + (A_1)^2 + (A_1)^2} = \sqrt{3}\,A_1$$

$$= \sqrt{(A_2)^2 + (A_2)^2 + (A_2)^2} = \sqrt{3}\,A_2$$

$$= \sqrt{(A_3)^2 + (A_3)^2 + (A_3)^2} = \sqrt{3}\,A_3$$

$$A_{oc2} = \sqrt{(A_1)^2 + (A_2)^2 + (-A_3)^2}$$

$$= \sqrt{(A_1)^2 + (A_1)^2 + (A_1)^2} = \sqrt{3}\,A_1$$

$$= \sqrt{(A_2)^2 + (A_2)^2 + (A_2)^2} \quad = \quad \sqrt{3}\ A_2$$

$$= \sqrt{(A_3)^2 + (A_3)^2 + (A_3)^2} \quad = \quad \sqrt{3}\ A_3$$

$$\cos\varphi_{12} = \frac{\mathbf{A_{oc1} \cdot A_{oc2}}}{A_{oc1}.\ A_{oc2}} = \frac{(A_1)^2}{\sqrt{3}\ A_1 .\ \sqrt{3}\ A_1}$$

$$= \frac{(v_x t_x R_x)^2}{\sqrt{3}(v_x t_x R_x).\sqrt{3}(v_x t_x R_x)} = \frac{1}{3} = 0.3333$$

Hence

$$\varphi_{12} = \cos^{-1}\left(\frac{(A_1)^2}{\sqrt{3}\ A_1 . \sqrt{3} A_1}\right) = \cos^{-1}\left(\frac{(A_2)^2}{\sqrt{3}\ A_2 . \sqrt{3} A_2}\right)$$

$$= \cos^{-1}\left(\frac{(A_3)^2}{\sqrt{3}\ A_3 . \sqrt{3} A_3}\right) = \cos^{-1}\left(\frac{1}{3}\right)$$

$$= \cos^{-1}(0.333) = 70.5287°$$

With the same method, these cosine angles in the six time-fields can be solved as follow:

$$\cos\varphi_{15} = \frac{\mathbf{A_{oc1} \cdot A_{oc5}}}{A_{oc1}.\ A_{oc5}} = \frac{(A_1)^2}{\sqrt{3}\ A_1 .\ \sqrt{3}\ A_1} = \frac{1}{3}$$

$$\cos\varphi_{26} = \frac{\mathbf{A_{oc2} \cdot A_{oc6}}}{A_{oc2}.\ A_{oc6}} = \frac{(A_1)^2}{\sqrt{3}\ A_1 .\ \sqrt{3}\ A_1} = \frac{1}{3}$$

$$\cos\varphi_{56} = \frac{\mathbf{A_{oc5} \cdot A_{oc6}}}{A_{oc5}.\ A_{oc6}} = \frac{(A_1)^2}{\sqrt{3}\ A_1 .\ \sqrt{3}\ A_1} = \frac{1}{3}$$

$$\cos\varphi_{34} = \frac{\mathbf{A_{oc3} \cdot A_{oc4}}}{A_{oc3}.\ A_{oc4}} = \frac{(A_1)^2}{\sqrt{3}\ A_1 .\ \sqrt{3}\ A_1} = \frac{1}{3}$$

$$\cos\varphi_{37} = \frac{\mathbf{A_{oc3} \cdot A_{oc7}}}{A_{oc3}.\ A_{oc7}} = \frac{(A_1)^2}{\sqrt{3}\ A_1 .\ \sqrt{3}\ A_1} = \frac{1}{3}$$

$$\cos\varphi_{48} = \frac{\mathbf{A_{oc4} \cdot A_{oc8}}}{A_{oc4}.\ A_{oc8}} = \frac{(A_1)^2}{\sqrt{3}\ A_1 .\ \sqrt{3}\ A_1} = \frac{1}{3}$$

$$\cos\varphi_{78} = \frac{\mathbf{A_{oc7}} \cdot \mathbf{A_{oc8}}}{A_{oc7}.\ A_{oc8}} = \frac{(A_1)^2}{\sqrt{3}\ A_1\ .\ \sqrt{3}\ A_1} = \frac{1}{3}$$

$$\cos\varphi_{14} = \frac{\mathbf{A_{oc1}} \cdot \mathbf{A_{oc4}}}{A_{oc1}.\ A_{oc4}} = \frac{(A_1)^2}{\sqrt{3}\ A_1\ .\ \sqrt{3}\ A_1} = \frac{1}{3}$$

$$\cos\varphi_{23} = \frac{\mathbf{A_{oc2}} \cdot \mathbf{A_{oc3}}}{A_{oc2}.\ A_{oc3}} = \frac{(A_1)^2}{\sqrt{3}\ A_1\ .\ \sqrt{3}\ A_1} = \frac{1}{3}$$

$$\cos\varphi_{58} = \frac{\mathbf{A_{oc5}} \cdot \mathbf{A_{oc8}}}{A_{oc5}.\ A_{oc8}} = \frac{(A_1)^2}{\sqrt{3}\ A_1\ .\ \sqrt{3}\ A_1} = \frac{1}{3}$$

$$\cos\varphi_{67} = \frac{\mathbf{A_{oc6}} \cdot \mathbf{A_{oc7}}}{A_{oc6}.\ A_{oc7}} = \frac{(A_1)^2}{\sqrt{3}\ A_1\ .\ \sqrt{3}\ A_1} = \frac{1}{3}$$

Conclusion:

$$\cos 70.5287^\circ = \cos\varphi_{12} = \cos\varphi_{14} = \cos\varphi_{15}$$
$$= \cos\varphi_{23} = \cos\varphi_{26} = \cos\varphi_{34}$$
$$= \cos\varphi_{37} = \cos\varphi_{48} = \cos\varphi_{56}$$
$$= \cos\varphi_{58} = \cos\varphi_{67} = \cos\varphi_{78}$$

9. To find the angle 54.7356°, for example, between $\mathbf{A}_{oc1} = A_1\mathbf{i} + A_2\mathbf{j} + A_3\mathbf{k}$ and $\mathbf{A}_{PAright} = 4A_1\mathbf{i} + 0\mathbf{j} + 0\mathbf{k}$, it can be solved as follow:

$$\mathbf{A}_{oc1}.\mathbf{A}_X = \mathbf{A}_{oc1}.\mathbf{A}_{PAright}$$

$$\mathbf{A}_{oc1}.\mathbf{A}_{PAright} = (A_1)\ (4A_1)+(A_2)\ (0)+(A_3)(0) = 4(A_1)^2 + 0 + 0$$

But in all cases in this chapter, $A_1 = A_2 = A_3$; as a result,

$$\mathbf{A}_{oc1}.\ \mathbf{A}_{PAright} = (A_1)(4A_1)+(A_2)\ (0)+(A_3)\ (0) = 4\ (A_1)^2$$
$$= 4\ (A_2)^2 = 4\ (A_3)^2$$

And these two magnitudes that are shown as symbols without boldness, A_{oc1}. $A_{PAright}$, are solved as follow:

$$A_{oc1} = \sqrt{(A_1)^2 + (A_2)^2 + (A_3)^2}$$

$$= \sqrt{(A_1)^2 + (A_1)^2 + (A_1)^2} = \sqrt{3}\,A_1$$

$$= \sqrt{3}\,A_2 = \sqrt{3}\,A_3$$

$$A_{PAright} = \sqrt{(4A_1)^2 + (0)^2 + (0)^2} = 4A_1$$

$$= 4A_2 = 4A_3$$

$$= \sqrt{(-4A_1)^2 + (0)^2 + (0)^2} = A_{PAleft}$$

$$= \sqrt{(0)^2 + (4A_1)^2 + (0)^2} = A_{PAupper}$$

$$= \sqrt{(0)^2 + (-4A_1)^2 + (0)^2} = A_{PAbottom}$$

$$= \sqrt{(0)^2 + (0)^2 + (4A_1)^2} = A_{PAfront}$$

$$= \sqrt{(0)^2 + (0)^2 + (-4A_1)^2} = A_{Paback}$$

$$\cos \gamma_{1X} = \frac{\mathbf{A_{oc1}.A_X}}{A_{oc1}.A_X} = \frac{\mathbf{A_{oc1}.A_{PAright}}}{A_{oc1}.A_{PAright}}$$

$$= \frac{4(A_1)^2}{\sqrt{3}\ A_1 .4\ A_1} = \frac{4(v_x t_x R_x)^2}{\sqrt{3}(v_x t_x R_x).4(v_x t_x R_x)}$$

$$= \frac{4(A_2)^2}{\sqrt{3}\ A_2 .4\ A_2} = \frac{4(v_y t_y R_y)^2}{\sqrt{3}(v_y t_y R_y).4(v_y t_y R_y)}$$

$$= \frac{4(A_3)^2}{\sqrt{3}\ A_3 .4\ A_3} = \frac{4(v_z t_z R_z)^2}{\sqrt{3}(v_z t_z R_z).4(v_z t_z R_z)}$$

$$= \frac{1}{\sqrt{3}} = 0.57735026918$$

Hence

$$\gamma_{1X} = \cos^{-1}\left(\frac{4(A_1)^2}{\sqrt{3}\,A_1\,.\,4A_1}\right) = \cos^{-1}\left(\frac{4(v_x t_x R_x)^2}{\sqrt{3}(v_x t_x R_x).4(v_x t_x R}\right.$$

$$= \cos^{-1}\left(\frac{4(A_2)^2}{\sqrt{3}\,A_2\,.\,4A_2}\right) = \cos^{-1}\left(\frac{4(v_y t_y R_y)^2}{\sqrt{3}(v_y t_y R_y).4(v_y t_y}\right.$$

$$= \cos^{-1}\left(\frac{4(A_3)^2}{\sqrt{3}\,A_3\,.\,4A_3}\right) = \cos^{-1}\left(\frac{4(v_z t_z R_z)^2}{\sqrt{3}(v_z t_z R_z).4(v_z t_z R}\right.$$

$$= \cos^{-1}\left(\frac{1}{\sqrt{3}}\right) = \cos^{-1}(0.57735026918)$$

$$= 54.7356°$$

With the same method, these cosine angles in the six time-fields can be solved as follow:

$$\cos\gamma_{2X} = \frac{A_{oc2}.A_{PAright}}{A_{oc2}.A_{PAright}} = \frac{4(A_1)^2}{\sqrt{3}\ A_1\ .\ 4\ A_1} = \frac{1}{\sqrt{3}}$$

$$\cos\gamma_{5X} = \frac{A_{oc5}.A_{PAright}}{A_{oc5}.A_{PAright}} = \frac{4(A_1)^2}{\sqrt{3}\ A_1\ .\ 4\ A_1} = \frac{1}{\sqrt{3}}$$

$$\cos\gamma_{6X} = \frac{A_{oc6}.A_{PAright}}{A_{oc6}.A_{PAright}} = \frac{4(A_1)^2}{\sqrt{3}\ A_1\ .\ 4\ A_1} = \frac{1}{\sqrt{3}}$$

$$\cos\gamma_{3(-X)} = \frac{A_{oc3}.A_{PAleft}}{A_{oc3}.A_{PAleft}} = \frac{4(A_1)^2}{\sqrt{3}\ A_1\ .\ 4\ A_1} = \frac{1}{\sqrt{3}}$$

$$\cos\gamma_{4(-X)} = \frac{A_{oc4}.A_{PAleft}}{A_{oc4}.A_{PAleft}} = \frac{4(A_1)^2}{\sqrt{3}\ A_1\ .\ 4\ A_1} = \frac{1}{\sqrt{3}}$$

$$\cos\gamma_{7(-X)} = \frac{A_{oc7}.A_{PAleft}}{A_{oc7}.A_{PAleft}} = \frac{4(A_1)^2}{\sqrt{3}\ A_1\ .\ 4\ A_1} = \frac{1}{\sqrt{3}}$$

$$\cos\gamma_{8(-X)} = \frac{A_{oc8}.A_{PAleft}}{A_{oc8}.A_{PAleft}} = \frac{4(A_1)^2}{\sqrt{3}\ A_1\ .\ 4\ A_1} = \frac{1}{\sqrt{3}}$$

$$\cos\gamma_{1Y} = \frac{A_{oc1}.A_{PAupper}}{A_{oc1}.A_{PAupper}} = \frac{4(A_1)^2}{\sqrt{3}\ A_1\ .\ 4\ A_1} = \frac{1}{\sqrt{3}}$$

$$\cos \gamma_{2Y} = \frac{A_{oc2} \cdot A_{PAupper}}{A_{oc2} . A_{PAupper}} = \frac{4(A_1)^2}{\sqrt{3}\ A_1 .\ 4\ A_1} = \frac{1}{\sqrt{3}}$$

$$\cos \gamma_{3Y} = \frac{A_{oc3} \cdot A_{PAupper}}{A_{oc3} . A_{PAupper}} = \frac{4(A_1)^2}{\sqrt{3}\ A_1 .\ 4\ A_1} = \frac{1}{\sqrt{3}}$$

$$\cos \gamma_{4Y} = \frac{A_{oc4} \cdot A_{PAupper}}{A_{oc4} . A_{PAupper}} = \frac{4(A_1)^2}{\sqrt{3}\ A_1 .\ 4\ A_1} = \frac{1}{\sqrt{3}}$$

$$\cos \gamma_{5(-Y)} = \frac{A_{oc5} \cdot A_{PAbottom}}{A_{oc5} . A_{PAbottom}} = \frac{4(A_1)^2}{\sqrt{3}\ A_1 .\ 4\ A_1} = \frac{1}{\sqrt{3}}$$

$$\cos \gamma_{6(-Y)} = \frac{A_{oc6} \cdot A_{PAbottom}}{A_{oc6} . A_{PAbottom}} = \frac{4(A_1)^2}{\sqrt{3}\ A_1 .\ 4\ A_1} = \frac{1}{\sqrt{3}}$$

$$\cos \gamma_{7(-Y)} = \frac{A_{oc7} \cdot A_{PAbottom}}{A_{oc7} . A_{PAbottom}} = \frac{4(A_1)^2}{\sqrt{3}\ A_1 .\ 4\ A_1} = \frac{1}{\sqrt{3}}$$

$$\cos \gamma_{8(-Y)} = \frac{A_{oc8} \cdot A_{PAbottom}}{A_{oc8} . A_{PAbottom}} = \frac{4(A_1)^2}{\sqrt{3}\ A_1 .\ 4\ A_1} = \frac{1}{\sqrt{3}}$$

$$\cos \gamma_{1Z} = \frac{A_{oc1} \cdot A_{PAfront}}{A_{oc1} . A_{PAfront}} = \frac{4(A_1)^2}{\sqrt{3}\ A_1 .\ 4\ A_1} = \frac{1}{\sqrt{3}}$$

$$\cos \gamma_{4Z} = \frac{A_{oc4} \cdot A_{PAfront}}{A_{oc4} . A_{PAfront}} = \frac{4(A_1)^2}{\sqrt{3}\ A_1 .\ 4\ A_1} = \frac{1}{\sqrt{3}}$$

$$\cos \gamma_{5Z} = \frac{A_{oc5} \cdot A_{PAfront}}{A_{oc5} . A_{PAfront}} = \frac{4(A_1)^2}{\sqrt{3}\ A_1 .\ 4\ A_1} = \frac{1}{\sqrt{3}}$$

$$\cos \gamma_{8Z} = \frac{A_{oc8} \cdot A_{PAfront}}{A_{oc8} . A_{PAfront}} = \frac{4(A_1)^2}{\sqrt{3}\ A_1 .\ 4\ A_1} = \frac{1}{\sqrt{3}}$$

$$\cos \gamma_{2(-Z)} = \frac{A_{oc2} \cdot A_{PAback}}{A_{oc2} . A_{PAback}} = \frac{4(A_1)^2}{\sqrt{3}\ A_1 .\ 4\ A_1} = \frac{1}{\sqrt{3}}$$

$$\cos \gamma_{3(-Z)} = \frac{A_{oc3} \cdot A_{PAback}}{A_{oc3} . A_{PAback}} = \frac{4(A_1)^2}{\sqrt{3}\ A_1 .\ 4\ A_1} = \frac{1}{\sqrt{3}}$$

$$\cos\gamma_{6(-Z)} = \frac{\mathbf{A_{oc6}.A_{PAback}}}{A_{oc6}.A_{PAback}} = \frac{4(A_1)^2}{\sqrt{3}\ A_1\ .\ 4\ A_1} = \frac{1}{\sqrt{3}}$$

$$\cos\gamma_{7(-Z)} = \frac{\mathbf{A_{oc7}.A_{PAback}}}{A_{oc7}.A_{PAback}} = \frac{4(A_1)^2}{\sqrt{3}\ A_1\ .\ 4\ A_1} = \frac{1}{\sqrt{3}}$$

Conclusion:

$\cos 54.7356^{\circ}$ = $\cos\gamma_{1X}$ = $\cos\gamma_{2X}$ = $\cos\gamma_{5X}$

Magic angle = $\cos\gamma_{6X}$ = $\cos\gamma_{3(-X)}$ = $\cos\gamma_{4(-X)}$

= $\cos\gamma_{7(-X)}$ = $\cos\gamma_{8(-X)}$ = $\cos\gamma_{1Y}$

= $\cos\gamma_{2Y}$ = $\cos\gamma_{3Y}$ = $\cos\gamma_{4Y}$

= $\cos\gamma_{5(-Y)}$ = $\cos\gamma_{6(-Y)}$ = $\cos\gamma_{7(-Y)}$

= $\cos\gamma_{8(-Y)}$ = $\cos\gamma_{1Z}$ = $\cos\gamma_{4Z}$

= $\cos\gamma_{5Z}$ = $\cos\gamma_{8Z}$ = $\cos\gamma_{2(-Z)}$

= $\cos\gamma_{3(-Z)}$ = $\cos\gamma_{6(-Z)}$ = $\cos\gamma_{7(-Z)}$

10. To find the angle 180°, for example, between $\mathbf{A_{oc1}} = A_1\mathbf{i} + A_2\mathbf{j} + A_3\mathbf{k}$ and $\mathbf{A_{oc7}} = -A_1\mathbf{i} - A_2\mathbf{j} - A_3\mathbf{k}$, it can be solved as follow:

$$\mathbf{A_{oc1}.A_{oc7}} = (A_1)(-A_1)+(A_2)(-A_2)+(A_3)(-A_3) = (-A_1)^2-(A_2)^2-(A_3)^2$$

But in all cases in this chapter, $A_1 = A_2 = A_3$; as a result,

$$\mathbf{A_{oc1}.A_{oc7}} = (A_1)(-A_1)+(A_2)(-A_2)+(A_3)(-A_3) = -(A_1)^2-(A_2)^2-(A_3)^2$$

$$= -(A_1)^2 - (A_1)^2 - (A_1)^2 = -3(A_1)^2$$

$$= -(A_2)^2 - (A_2)^2 - (A_2)^2 = -3(A_2)^2$$

$$= -(A_3)^2 - (A_3)^2 - (A_3)^2 = -3(A_3)^2$$

And these two magnitudes that are shown as symbols without boldness, $A_{oc1}.A_{oc7}$, are solved as follow:

$$A_{oc1} = \sqrt{(A_1)^2 + (A_2)^2 + (A_3)^2}$$

$$= \sqrt{(A_1)^2 + (A_1)^2 + (A_1)^2} = \sqrt{3}\,A_1$$

$$= \sqrt{(A_2)^2 + (A_2)^2 + (A_2)^2} = \sqrt{3}\,A_2$$

$$= \sqrt{(A_3)^2 + (A_3)^2 + (A_3)^2} = \sqrt{3}\,A_3$$

$$A_{oc7} = \sqrt{(-A_1)^2 + (-A_2)^2 + (-A_3}$$

$$= \sqrt{(A_1)^2 + (A_1)^2 + (A_1)^2} = \sqrt{3}\,A_1$$

$$= \sqrt{(A_2)^2 + (A_2)^2 + (A_2)^2} = \sqrt{3}\,A_2$$

$$= \sqrt{(A_3)^2 + (A_3)^2 + (A_3)^2} = \sqrt{3}\,A_3$$

$$\cos\delta_{17} = \frac{\mathbf{A_{oc1} \cdot A_{oc7}}}{A_{oc1}.\ A_{oc7}} = \frac{-3\,(A_1)^2}{\sqrt{3}\ A_1\,.\ \sqrt{3}\,A_1}$$

$$= \frac{-3\,(v_x t_x R_x)^2}{\sqrt{3}(v_x t_x R_x).\sqrt{3}(v_x t_x R_x)} = -1$$

Hence

$$\delta_{17} = \cos^{-1}\left(\frac{-3\,(A_1)^2}{\sqrt{3}\,A_1\,.\,\sqrt{3}A_1}\right) = \cos^{-1}\left(\frac{-3\,(A_2)^2}{\sqrt{3}\,A_2\,.\,\sqrt{3}A_2}\right)$$

$$= \cos^{-1}\left(\frac{-3\,(A_3)^2}{\sqrt{3}\,A_3\,.\,\sqrt{3}A_3}\right) = \cos^{-1}\left(\frac{-3}{3}\right)$$

$$= \cos^{-1}(-1) = 180°$$

With the same method, these cosine angles in the six time-fields can be solved as follow:

$$\cos\delta_{28} = \frac{\mathbf{A_{oc2} \cdot A_{oc8}}}{A_{oc2.}\ A_{oc8}} = \frac{-3\,(A_1)^2}{\sqrt{3}\ A_1\,.\ \sqrt{3}\,A_1} = 1$$

$$\cos\delta_{35} = \frac{\mathbf{A_{oc3} \cdot A_{oc5}}}{A_{oc3.}\ A_{oc5}} = \frac{-3\,(A_1)^2}{\sqrt{3}\ A_1\,.\ \sqrt{3}\,A_1} = 1$$

$$\cos \delta_{46} = \frac{\mathbf{A}_{oc4}.\mathbf{A}_{oc6}}{A_{oc4}.\ A_{oc6}} = \frac{-3\ (A_1)^2}{\sqrt{3}\ A_1\ .\ \sqrt{3}\ A_1} = 1$$

Conclusion:

$$\cos 180^\circ = \cos \delta_{17} = \cos \delta_{28} = \cos \delta_{35}$$

$$= \cos \delta_{46}$$

11. To find the edge of the time-field, it can be solved as follow:

In all cases in this chapter, $A_1 = A_2 = A_3$; as a result,

One of the edges of the time-fields is the magnitude of $\mathbf{A}_{oc1}$. It can be solved as follow:

A_{oc1} = One of the edges of the time-fields

$$= \sqrt{(A_1)^2 + (A_2)^2 + (A_3)^2}$$

$$= \sqrt{(A_1)^2 + (A_1)^2 + (A_1)^2} = \sqrt{3}\ A_1$$

$$= \sqrt{(A_2)^2 + (A_2)^2 + (A_2)^2} = \sqrt{3}\ A_2$$

$$= \sqrt{(A_3)^2 + (A_3)^2 + (A_3)^2} = \sqrt{3}\ A_3$$

12. To find one side of the four sides of the base of the time-field, symbol b_q, it can be solved as follow:

$b_q = \sqrt{c^2}$ = a quarter of the whole length of the base

But $\sqrt{c^2} = \sqrt{a^2 + b^2}$

$a = (d_h) = \sqrt{2}\ (A_1)$

$a^2 = (d_h)^2 \quad = \quad 2\,(A_1)^2$

$b = (d_h) \quad = \quad \sqrt{2}\,(A_1)$

$b^2 = (d_h)^2 \quad = \quad 2\,(A_1)^2$

$b_q = \sqrt{c^2} \quad = \quad \sqrt{2(A_1)^2 + 2(A_1)^2}$

$= \sqrt{4(A_1)^2} \quad = \quad 2\,A_1$

$= 2\,A_2 \quad = \quad 2\,A_3$

13. To find a half of the diagonal at the base of each time-field (d_h), it can be solved as follow:

$d_h = \sqrt{a^2}$ = a half of the diagonal at the base of each time-field

But $\sqrt{a^2} = \sqrt{c^2 - b^2}$

$c = A_{oc1} \quad = \quad \sqrt{(A_1)^2 + (A_2)^2 + (A_3)^2}$

$= \sqrt{3}\,(A_1) \quad = \quad 1.732\,A_1$

$c^2 = 3\,(A_1)^2$

$b = d_h \quad = \quad \sqrt{c^2 - a^2}$

$= \sqrt{3(A_1)^2 - (A_{\cdot}} \quad = \quad \sqrt{2}\,(A_1)$

$b^2 = (d_h)^2 \quad = \quad 2\,(A_1)^2$

$$d_h = \sqrt{a^2} = \sqrt{3(A_1)^2 - 2(A_1)^2}$$

$$= \sqrt{(A_1)^2} = A_1$$

$$= A_2 = A_3$$

References

"The Poverty of Economics" by Robert Kuttner, The Atlantic Monthly, February 1985. (2016). *UNZ.org*. Retrieved 7 September 2016, from http://www.unz.org/Pub/AtlanticMonthly-1985feb-00074

10 Greatest Unsolved Mysteries In Physics. (2016). *IFLScience*. Retrieved 12 October 2016, from http://www.iflscience.com/physics/greatest-mysteries-physics/

A Brief History of Time. (2016). *En.wikipedia.org*. Retrieved 25 October 2016, from https://en.wikipedia.org/wiki/A_Brief_History_of_Time

Akins, W. (1980). *The secret power of the pyramids* (1st ed.). New York: Watts.

Albert A. Michelson - Biographical. (2016). *Nobelprize.org*. Retrieved 24 October 2016, from https://www.nobelprize.org/nobel_prizes/physics/laureates/1907/michelson-bio.html

Albert Einstein | German-American physicist. (2016). *Encyclopedia Britannica*. Retrieved 24 October 2016, from https://global.britannica.com/biography/Albert-Einstein

All About Magnets. (2016). *Hometrainingtools.com*. Retrieved 13 June 2016, from http://www.hometrainingtools.com/a/all-about-magnets

Angle. *En.wikipedia.org*. Retrieved 5 November 2016, from https://en.wikipedia.org/wiki/Angle

Arthur Eddington facts, information, pictures | Encyclopedia.com articles about Arthur Eddington. (2016). *Encyclopedia.com*. Retrieved 24 October 2016, from http://www.encyclopedia.com/people/history/historians-miscellaneous-biographies/arthur-eddington

ASP: How Fast Are You Moving When You Are Sitting Still?. (2016). *Astrosociety.org*. Retrieved 24 October 2016, from https://astrosociety.org/edu/publications/tnl/71/howfast.html

Boonsri, P. (2000). *Pyramid: The Solved Mystery*. Bangkok: Concept Line Co., Ltd.

Boonsri, P. (2002). Mathematics proof of time dimension as a vector. *The Journal Of The Royal Institute Of Thailand*, *27*(2), 399-407.

Boonsri, P. (2003). *Geometry of time dimension*. Presentation, Songkla.

Boonsri, P. (2006). *The Mathematical Proof of the First Serious Mistake in Mainstream Price Microeconomics Theory, MPMT*. Presentation, All China Economic International Conference at City University of Hong Kong. 18-20 December 2006.

Boonsri, P. (2006). *The Mathematical Proof of the Second Serious Mistake in Mainstream Price Microeconomic Theory, MPMT*. Presentation, All China Economic International Conference at City University of Hong Kong. 18-20 December 2006.

Boonsri, P. (2007). *Some ways to correct 2 mathematically Proven Serious Mistakes in Mainstream Price Microeconomic Theory, MPMT*. Presentation, Second All China Economic

International Conference at City University of Hong Kong. 12-14 December 2007.

Boonsri, P. (2011). The inconsistency of the definition of demand and supply in price microeconomics Theory and some correction. In *Proceeding of The 2nd BSRU International Conference 2011* (pp. 297-304). Bangkok: Bansomdejchaopraya Rajabhat University.

Boonsri, P. (2012). The Time Equation. In *Proceeding of the Research for Thai Sustainable Development* (pp. 28-35). Bangkok: Bansomdejchaopraya Rajabhat University.

Boonsri, P. (2013). *The Time Equation and the Weak Points of the Theoretical Demand and the Supply Curves in Microeconomic Theory. The Journal of Interdisciplinary Networks*. Retrieved 12 June 2016, from http://www.icird.mju.ac.th/icird2013/JIN1_CD.pdf

Boonsri, P. (2014). The study on time to develop an economic theory. In *The 5th International Conference on Business and economics* (pp. 186-197). Bangkok: UP Organizer and Publication Co., Ltd.

Boonsri, P. (2014). The time as the linkage between microeconomics and macroeconomics. In *The 3rd International Conference on Production and Supply Chain Management* (pp. 329-343). Newcastle Upon Type (Type And Wear): Cambridge Scholars Publishing Ltd.

Boonsri, P. (2015). Keynes' theory with time as the linkage between micro- and macroeconomics. In *The International Conference on management, business, and economics & The 3rd*

International Conference on Tourism, Transport, and Logistics (pp. 227-241). Bangkok: Sripatum University Press.

Boonsri, P. (2016). *Scientific Evidence of Existence of Time Dimension in Three-Dimensional Body*. San Francisco, California: https://kku.academia.edu/PoramestBoonsri.

Boonsri, P. (2016). *The Research on Linear Time Equation and the linkage to the Theory of Everything*. Bangkok: BSRU.

boson | subatomic particle. (2016). *Encyclopedia Britannica*. Retrieved 24 October 2016, from https://global.britannica.com/science/boson

Bydder, M., Rahal, A., Fullerton, G., & Bydder, G. (2007). The magic angle effect: A source of artifact, determinant of image contrast, and technique for imaging. *J. Magn. Reson. Imaging, 25*(2), 290-300. http://dx.doi.org/10.1002/jmri.20850

Cartesian coordinate system. (2016). *Wikipedia*. Retrieved 11 June 2016, from http://en.wikipedia.org/wiki/Cartesian_coordinate_system

Charpentier, T., Fermon, C., & Virlet, J. (1998). Numerical and theoretical analysis of multiquantum magic-angle spinning experiments. *The Journal Of Chemical Physics, 109*(8), 3116. http://dx.doi.org/10.1063/1.476903

Cobweb model. En.wikipedia.org. Retrieved 29 October 2016, from https://en.wikipedia.org/wiki/Cobweb_model

COMMITTEE MEMBERS OF THE CHAIPATTANA FOUNDATION. (2016). *Chaipat.or.th*. Retrieved 29 November 2016, from http://www.chaipat.or.th/chaipat_english/index.php?option=co

m_content&view=article&id=4107:committee-members&catid=717:about-us&Itemid=304

Davies, P. (1995). *About time*. New York: Simon & Schuster.

Demand. (2016). *En.wikipedia.org*. Retrieved 23 November 2016, from https://en.wikipedia.org/wiki/Demand

Disney buys Pixar - Jan. 25, 2006. (2016). *Money.cnn.com*. Retrieved 25 October 2016, from http://money.cnn.com/2006/01/24/news/companies/disney_pixar_deal/

Drucker, P. (1981). Toward the Next Economics. In D. Bell & I. Kristol, *The Crisis in Economic Theory*(1st ed., pp. 4-18). New York: Basic Books.

Drucker, P. (1981). *Toward the next economics, and other essays*. New York: Harper & Row.

Earth's magnetic field. (2016). *Wikipedia*. Retrieved 11 June 2016, from http://en.wikipedia.org/wiki/Earth%27s_magnetic_field

Earth's orbit. (2016). *Wikipedia*. Retrieved 11 June 2016, from https://en.wikipedia.org/wiki/Earth%27s_orbit

Edward Morley - New World Encyclopedia. (2016). *Newworldencyclopedia.org*. Retrieved 24 October 2016, from http://www.newworldencyclopedia.org/entry/Edward_Morley

Eisinger, J. (2016). Thoughts on Einstein and general relativity. *Physics Today*, *69*(10), 12-12. http://dx.doi.org/10.1063/pt.3.3312

electromagnetic field | physics. (2015). *Encyclopedia Britannica*. Retrieved 24 October 2016, from https://global.britannica.com/science/electromagnetic-field

Electromagnetic theory. (2016). *TheFreeDictionary.com*. Retrieved 24 October 2016, from http://encyclopedia2.thefreedictionary.com/Electromagnetic+theory

Energy Development. (2016). *Chaipat.or.th*. Retrieved 3 December 2016, from http://www.chaipat.or.th/chaipat_english/index.php?option=com_content&view=article&id=4146:development&catid=722:our-work&Itemid=300

Enrico Fermi - Facts. (2016). *Nobelprize.org*. Retrieved 24 October 2016, from https://www.nobelprize.org/nobel_prizes/physics/laureates/1938/fermi-facts.html

Enrico Fermi. (2016). *Wikipedia*. Retrieved 11 June 2016, from http://en.wikipedia.org/wiki/Enrico_Fermi

Ernest Lawrence - Biographical. (2016). *Nobelprize.org*. Retrieved 24 October 2016, from https://www.nobelprize.org/nobel_prizes/physics/laureates/1939/lawrence-bio.html

Ernest Walton - Dictionary definition of Ernest Walton | Encyclopedia.com: FREE online dictionary. (2016). *Encyclopedia.com*. Retrieved 25 October 2016, from http://www.encyclopedia.com/history/encyclopedias-almanacs-transcripts-and-maps/ernest-walton

Erwin Schrodinger facts, information, pictures | Encyclopedia.com articles about Erwin Schrodinger. (2016). *Encyclopedia.com.* Retrieved 24 October 2016, from http://www.encyclopedia.com/people/science-and-technology/physics-biographies/erwin-schrodinger

Experimental and theoretical evidence for the magic angle in transmission electron energy loss spectroscopy. (2016). *Sciencedirect.com.* Retrieved 12 June 2016, from http://www.sciencedirect.com/science/article/pii/S0304399103 00113X

Four-dimensional space. En.wikipedia.org. Retrieved 8 November 2016, from https://en.wikipedia.org/wiki/Four-dimensional_space

Fritjof Capra. (2016). *En.wikipedia.org.* Retrieved 6 December 2016, from https://en.wikipedia.org/wiki/Fritjof_Capra

Fullbrook, E. (2004). *A guide to what's wrong with economics* (1st ed.). London: Anthem.

Gaillard, F. (2016). *Magic angle effect - MRI artifact | Radiology Reference Article | Radiopaedia.org. Radiopaedia.org.* Retrieved 24 October 2016, from https://radiopaedia.org/articles/magic-angle-effect-mri-artifact

Galileo Galilei - New World Encyclopedia. (2016). *Newworldencyclopedia.org.* Retrieved 25 October 2016, from http://www.newworldencyclopedia.org/entry/Galileo_Galilei

general theory of relativity Facts, information, pictures | Encyclopedia.com articles about general theory of relativity.

(2016). *Encyclopedia.com*. Retrieved 5 September 2016, from http://www.encyclopedia.com/topic/general_theory_of_relativity.aspx

geometry facts, information, pictures | Encyclopedia.com articles about geometry. *Encyclopedia.com*. Retrieved 8 November 2016, from http://www.encyclopedia.com/science-and-technology/mathematics/mathematics/geometry

Great Depression in the United States. (2016). *En.wikipedia.org*. Retrieved 28 November 2016, from https://en.wikipedia.org/wiki/Great_Depression_in_the_United_States

hadron. (2016). *TheFreeDictionary.com*. Retrieved 24 October 2016, from http://www.thefreedictionary.com/hadron

Hadrons - DiracDelta Science & Engineering Encyclopedia. (2016). *Diracdelta.co.uk*. Retrieved 25 October 2016, from http://www.diracdelta.co.uk/science/source/h/a/hadrons/source.html#.WA-DnPmLTIU

Hawking, S. (1998). *A brief history of time*. New York: Bantam Books.

Heilbroner, R. & Milberg, W. (1995). *The crisis of vision in modern economic thought* (1st ed.). Cambridge [England]: Cambridge University Press.

Henri Poincaré - New World Encyclopedia. (2016). *Newworldencyclopedia.org*. Retrieved 25 October 2016, from http://www.newworldencyclopedia.org/entry/Henri_Poincaré

Hermann Minkowski | German mathematician. (2016). *Encyclopedia Britannica*. Retrieved 12 September 2016, from https://global.britannica.com/biography/Hermann-Minkowski

Hermann Minkowski Facts, information, pictures | Encyclopedia.com articles about Hermann Minkowski. (2016). *Encyclopedia.com*. Retrieved 5 September 2016, from http://www.encyclopedia.com/topic/Hermann_Minkowski.aspx

Higher-dimensional geometry - Encyclopedia of Mathematics. *Encyclopediaofmath.org*. Retrieved 8 November 2016, from https://www.encyclopediaofmath.org/index.php/Higher-dimensional_geometry

History. (2016). *Ministry of Agriculture and Cooperatives*. Retrieved 29 November 2016, from https://eng.moac.go.th/main.php?filename=History

Hoffman, J. (2014). *Robert Pound, Physicist Whose Work Advanced Medicine, Is Dead at 90*. *Nytimes.com*. Retrieved 25 October 2016, from http://www.nytimes.com/2010/04/20/us/20pound.html?_r=0

How "Fast" is the Speed of Light?. (2016). *Grc.nasa.gov*. Retrieved 24 October 2016, from https://www.grc.nasa.gov/www/k-12/Numbers/Math/Mathematical_Thinking/how_fast_is_the_speed.htm

HubbleSite: Black Holes: Gravity's Relentless Pull interactive: Encyclopedia. (2005). *Hubblesite.org*. Retrieved 27 October 2016, from http://hubblesite.org/explore_astronomy/black_holes/encyc_mod3_q4.html

Hyperinflation. (2016). *En.wikipedia.org*. Retrieved 28 November 2016, from https://en.wikipedia.org/wiki/Hyperinflation

Isaac Newton | Biography, Facts, Discoveries, Laws, & Inventions. (2016). *Encyclopedia Britannica*. Retrieved 3 September 2016, from https://global.britannica.com/biography/Isaac-Newton

James Chadwick - New World Encyclopedia. (2016). *Newworldencyclopedia.org*. Retrieved 24 October 2016, from http://www.newworldencyclopedia.org/entry/James_Chadwick

James Clerk Maxwell - New World Encyclopedia. (2016). *Newworldencyclopedia.org*. Retrieved 24 October 2016, from http://www.newworldencyclopedia.org/entry/James_Clerk_Maxwell

John Cockcroft - Facts. (2016). *Nobelprize.org*. Retrieved 24 October 2016, from https://www.nobelprize.org/nobel_prizes/physics/laureates/1951/cockcroft-facts.html

John Cockcroft. *En.wikipedia.org*. Retrieved 29 October 2016, from https://en.wikipedia.org/wiki/John_Cockcroft

John Hicks. (2016). *En.wikipedia.org*. Retrieved 28 November 2016, from https://en.wikipedia.org/wiki/John_Hicks

John Maynard Keynes | British economist. (2016). *Encyclopedia Britannica*. Retrieved 28 November 2016, from https://global.britannica.com/biography/John-Maynard-Keynes

John Maynard Keynes. (2016). *En.wikipedia.org*. Retrieved 28 November 2016, from https://en.wikipedia.org/wiki/John_Maynard_Keynes

John Richard Hicks - New World Encyclopedia. (2016). *Newworldencyclopedia.org*. Retrieved 28 November 2016, from http://www.newworldencyclopedia.org/entry/John_Richard_Hicks

Joseph E. Stiglitz | American economist. *Encyclopedia Britannica*. Retrieved 23 November 2016, from https://global.britannica.com/biography/Joseph-Stiglitz

Joseph Stiglitz. (2016). *En.wikipedia.org*. Retrieved 23 November 2016, from https://en.wikipedia.org/wiki/Joseph_Stiglitz

Jules Henri Poincare facts, information, pictures | Encyclopedia.com articles about Jules Henri Poincare. (2016). *Encyclopedia.com*. Retrieved 24 October 2016, from http://www.encyclopedia.com/people/science-and-technology/mathematics-biographies/jules-henri-poincare

Kaku, M. & Trainer, J. (1987). *Beyond Einstein* (pp. 26-30). New York: Bantam Books.

Kaku, M. (2016). *The One-inch Equation to Explain All Physical Laws*. *Bigthink.com*. Retrieved 9 September 2016, from http://bigthink.com/dr-kakus-universe/the-one-inch-equation-to-explain-all-physical-laws

Karl Marx | German philosopher. (2016). *Encyclopedia Britannica*. Retrieved 28 November 2016, from https://global.britannica.com/biography/Karl-Marx

Karl Marx. (2016). *Wikipedia*. Retrieved 28 November 2016, from https://simple.wikipedia.org/wiki/Karl_Marx

Kassakorn, N. (2014). The Synthesis of the Researches on the "Holistic Thinking Model" to develop the Logistics and Supply Chain from the Royal Speech of His Majesty, Bhumibol Adulyadej the Great. In:. In *The 2nd International Conference on Tourism, Transport, and Logistics* (pp. 333-344). Bangkok: UP.

light Facts, information, pictures | Encyclopedia.com articles about light. (2016). *Encyclopedia.com*. Retrieved 5 September 2016, from http://www.encyclopedia.com/topic/light.aspx

Louis de Broglie - Biography, Facts and Pictures. (2016). *Famousscientists.org*. Retrieved 24 October 2016, from http://www.famousscientists.org/louis-de-broglie/

Louis Victor de Broglie facts, information, pictures | Encyclopedia.com articles about Louis Victor de Broglie. (2016). *Encyclopedia.com*. Retrieved 24 October 2016, from http://www.encyclopedia.com/people/science-and-technology/chemistry-biographies/louis-victor-de-broglie

Luis Walter Alvarez. (2016). *En.wikipedia.org*. Retrieved 24 November 2016, from https://en.wikipedia.org/wiki/Luis_Walter_Alvarez

Luis Walter Alvarez. (2016). *Wikipedia*. Retrieved 11 June 2016, from https://en.wikipedia.org/wiki/Luis_Walter_Alvarez

Magic angle (EELS). (2016). *Wikipedia*. Retrieved 11 June 2016, from https://en.wikipedia.org/wiki/Magic_angle_(EELS)

Magnetic field. (2016). *Wikipedia*. Retrieved 11 June 2016, from http://en.wikipedia.org/wiki/Magnetic_field

magnetic force | physics. (2016). *Encyclopedia Britannica*. Retrieved 24 October 2016, from https://global.britannica.com/science/magnetic-force

Market demand schedule. (2016). *En.wikipedia.org*. Retrieved 23 November 2016, from https://en.wikipedia.org/wiki/Market_demand_schedule

Markosian, N. (2002). *Time. Plato.stanford.edu*. Retrieved 24 October 2016, from http://plato.stanford.edu/entries/time/

Matthews, P. (1998). *Vector calculus*. Berlin: Springer.

Max Born - Biographical. (2016). *Nobelprize.org*. Retrieved 24 October 2016, from http://www.nobelprize.org/nobel_prizes/physics/laureates/1954/born-bio.html

Max Planck - Biographical. (2016). *Nobelprize.org*. Retrieved 25 October 2016, from https://www.nobelprize.org/nobel_prizes/physics/laureates/1918/planck-bio.html

Mendell, H. (2004). *Aristotle and Mathematics. Plato.stanford.edu*. Retrieved 25 October 2016, from http://plato.stanford.edu/entries/aristotle-mathematics/

Missions and Operating Principles. (2016). *Chaipat.or.th*. Retrieved 29 November 2016, from http://www.chaipat.or.th/chaipat_english/index.php?option=com_content&view=article&id=4120:missions-and-operating-principles&catid=717:about-us&Itemid=295

Murashige, T. & Skoog, F. (1962). *T. Murashige and F. Skoog, "A Revised Medium for Rapid Growth and Bioassays with Tobacco Tissue Cultures," Physiologia Plantarum, Vol. 15, No. 3, 1962, pp. 473-497. http://dx.doi.org/10.1111/j.1399-3054.1962.tb08052.x - References - Scientific Research Publish. Scirp.org.* Retrieved 28 November 2016, from http://www.scirp.org/(S(vtj3fa45qm1ean45vvffcz55))/reference/ReferencesPapers.aspx?ReferenceID=1079764

Niels Bohr - New World Encyclopedia. (2016). *Newworldencyclopedia.org.* Retrieved 24 October 2016, from http://www.newworldencyclopedia.org/entry/Niels_Bohr

Nuclear Magnetic Resonance Facts, information, pictures | Encyclopedia.com articles about Nuclear Magnetic Resonance. (2016). *Encyclopedia.com.* Retrieved 5 September 2016, from http://www.encyclopedia.com/topic/Nuclear_Magnetic_Resonance.aspx

nuclear reactor Facts, information, pictures | Encyclopedia.com articles about nuclear reactor. (2016). *Encyclopedia.com.* Retrieved 5 September 2016, from http://www.encyclopedia.com/topic/nuclear_reactor.aspx

ORDPB,. (2016). *The developer King-Office of The Royal Development Projects Board (ORDPB). 61.91.22.74.*

Retrieved 3 December 2016, from http://61.91.22.74/Rdpbweb2016/King

Otto Hahn, Lise Meitner, and Fritz Strassmann | Chemical Heritage Foundation. (2016). *Chemheritage.org*. Retrieved 25 October 2016, from https://www.chemheritage.org/historical-profile/otto-hahn-lise-meitner-and-fritz-strassmann

Otto Hahn. (2016). *Wikipedia*. Retrieved 11 June 2016, from http://en.wikipedia.org/wiki/Otto_Hahn

Parker, B. (1987). *Search for a Super Theory from Atom to Superstrings*. New York: Plenum Press.

Pashigian, P. (2008). *cobweb theorem : The New Palgrave Dictionary of Economics*. *Dictionaryofeconomics.com*. Retrieved 29 October 2016, from http://www.dictionaryofeconomics.com/article?id=pde2008_C000186

Paul Adrien Maurice Dirac facts, information, pictures | Encyclopedia.com articles about Paul Adrien Maurice Dirac. (2016). *Encyclopedia.com*. Retrieved 25 October 2016, from http://www.encyclopedia.com/people/science-and-technology/physics-biographies/paul-adrien-maurice-dirac

Peter Drucker - New World Encyclopedia. *Newworldencyclopedia.org*. Retrieved 13 November 2016, from http://www.newworldencyclopedia.org/entry/Peter_Drucker

Pixar Animation Studios - Dictionary definition of Pixar Animation Studios | Encyclopedia.com: FREE online dictionary. (2016). *Encyclopedia.com*. Retrieved 25 October 2016, from http://www.encyclopedia.com/books/politics-and-business-magazines/pixar-animation-studios

Point (geometry). (2016). *Wikipedia*. Retrieved 11 June 2016, from http://en.wikipedia.org/wiki/Point_%28geometry%29

Point (geometry). *En.wikipedia.org*. Retrieved 8 November 2016, from https://en.wikipedia.org/wiki/Point_(geometry)

Projects initiated by His Majestic King Rama IX, the Great. (2016). *Office of the Royal Development Projects Board*. Retrieved 29 November 2016, from http://www.rdpb.go.th/rdpb/Front/Projects/ImportantDetail.aspx?projectid=41

Pyramid power. (2016). *En.wikipedia.org*. Retrieved 24 November 2016, from https://en.wikipedia.org/wiki/Pyramid_power

Pyramid power. (2016). *Wikipedia*. Retrieved 12 September 2016, from https://en.wikipedia.org/wiki/Pyramid_power

quantum mechanics. (2016). *TheFreeDictionary.com*. Retrieved 24 October 2016, from http://encyclopedia2.thefreedictionary.com/quantum+mechanics

Quartz crisis. *En.wikipedia.org*. Retrieved 30 October 2016, from https://en.wikipedia.org/wiki/Quartz_crisis

RAZOR SHARPENING PYRAMID by Brazdadj. (2016). *YouTube*. Retrieved 24 November 2016, from https://www.youtube.com/watch?v=vMaFmuxC9Og

Robert J. Van de Graaff. (2016). *Wikipedia*. Retrieved 11 June 2016, from http://en.wikipedia.org/wiki/Robert_J._Van_de_Graaff

Ruffell, A. (1981). *Pyramid power* (1st ed.). London: J. MacRae Books.

Rutherford, Ernest – Biography – Te Ara Encyclopedia of New Zealand. (2016). *Teara.govt.nz*. Retrieved 24 October 2016, from http://www.teara.govt.nz/en/biographies/3r37/rutherford-ernest

Scientists, T. & Scientists, L. (2016). *Paul Dirac - Biography, Facts and Pictures*. *Famousscientists.org*. Retrieved 24 October 2016, from http://www.famousscientists.org/paul-dirac/

Sir Joseph John Thomson - Dictionary definition of Sir Joseph John Thomson | Encyclopedia.com: FREE online dictionary. (2016). *Encyclopedia.com*. Retrieved 24 October 2016, from http://www.encyclopedia.com/science/encyclopedias-almanacs-transcripts-and-maps/sir-joseph-john-thomson

Sivapraphagorn, A. (2006). *Effect of Time Field on Rate of Redox Reaction. Journal of Science and technology, Ubon Rajathani University,*. Retrieved 12 June 2016, from http://www.ubu.ac.th/web/files_up/08f2013032215272079.pdf

Sivapraphagorn, A., Boonsri, P., & Samran, S. (2014). *Feasibility of an Electromagnetic Cell. http://www.ubu.ac.th/web/content.php?content=0800001596*. Retrieved 16 September 2016, from http://www.ubu.ac.th/web/files_up/08f2015010515575267.pdf

Sivapraphagorn, A., Boonsri, P., & Samran, S. (2014). *Feasibility of an Electromagnetic Cell. Journal of Science and Technology, Ubon Ratchathani University*. Retrieved 25 November 2016, from http://www.ubu.ac.th/web/files_up/08f2015010515575267.pdf

Sivapraphagorn, A., Samran, S., Pimmongkol, A., & Parasarn, W. (2011). *Possibility of Activated Time Dimension Effects on Rate of Redox Reaction and Growth Rate of Tissue Culture Plants. http://kasetsartjnatsci.kasetsart.org/.* Retrieved 16 September 2016, from http://kasetsartjournal.ku.ac.th/kuj_files/2011/A110301111735 3593.pdf

Space-time - Encyclopedia of Mathematics. (2016). *Encyclopediaofmath.org.* Retrieved 25 October 2016, from https://www.encyclopediaofmath.org/index.php/Space-time

special relativity | physics. (2016). *Encyclopedia Britannica.* Retrieved 12 September 2016, from https://global.britannica.com/science/special-relativity

StateMaster - Encyclopedia: Pieter Zeeman. (2016). *Statemaster.com.* Retrieved 24 October 2016, from http://www.statemaster.com/encyclopedia/Pieter-Zeeman

supply and demand | economics. Encyclopedia Britannica. Retrieved 23 November 2016, from https://global.britannica.com/topic/supply-and-demand

supply and demand facts, information, pictures | Encyclopedia.com articles about supply and demand. Encyclopedia.com. Retrieved 22 November 2016, from http://www.encyclopedia.com/social-sciences-and-law/economics-business-and-labor/economics-terms-and-concepts/supply-and-demand

The First Cyclotrons - Ernest Lawrence and the Cyclotron: AIP History Center Web Exhibit. (2016). *Aip.org*. Retrieved 24 October 2016, from https://www.aip.org/history/exhibits/lawrence/first.htm

THE NATURAL PHENOMENA OF ANTIGRAVITATION AND INVISIBILITY IN INSECTS AND THE GREBENNIKOV CAVERNOUS STRUCTURES EFFECT. (2016). *Keelynet.com*. Retrieved 13 June 2016, from http://www.keelynet.com/greb/greb.htm

Time | Internet Encyclopedia of Philosophy. (2016). *Iep.utm.edu*. Retrieved 24 October 2016, from http://www.iep.utm.edu/time/

Toth, M. & Nielsen, G. (1985). *Pyramid power* (1st ed.). New York: Destiny Books.

Veblen, T. (1898). Why is Economics not an Evolutionary Science?. *The Quarterly Journal Of Economics*, *12*(4 (July, 1898), 373-397. Retrieved from http://www.jstor.org/stable/1882952

Werner Heisenberg - Biographical. (2016). *Nobelprize.org*. Retrieved 24 October 2016, from http://www.nobelprize.org/nobel_prizes/physics/laureates/1932/heisenberg-bio.html

Willes, M. (1981). Rational Expectations" as a Counterrevolution. In D. Bell & I. Kristol, *The Crisis in Economic Theory* (1st ed., pp. 85-90). New York: Basic Books.

Willes, P. (1983). Ideology, Methodology, and Neoclassical Economics. In A. Eichner, *Why Economics is Not Yet a Science* (1st ed., pp. 41-89). New York: M.E. Sharpe.

Wolfgang Pauli - Biographical. (2016). *Nobelprize.org.* Retrieved 24 October 2016, from http://www.nobelprize.org/nobel_prizes/physics/laureates/1945/pauli-bio.html

Wright, D. & New, B. (1992). *Essential calculus with applications.* Lexington, Mass.: D.C. Heath.

Zeeman, Pieter (1865-1943). (2016). *Daviddarling.info.* Retrieved 25 October 2016, from http://www.daviddarling.info/encyclopedia/Z/Zeeman.html

About the Author

Associate Professor Dr. Poramest Boonsri

Poramest Boonsri was born and raised in Bangkok, Thailand and still lives in the thriving city today, with his wife and daughter.

He attended the prestigious Faculty of Science at Khon Khen University, before going on to Ramkhamhaeng University where he attained a second-class honour in quantitative economics. Poramest has also obtained a master's degree in quantitative economics, an Australian's Diploma in Interdisciplinary Study and a PhD, the last coming from his research on a new price model in economics from Bansomdejchaoprava Rajabhat University, one of the oldest universities of Thailand.

Poramest has been an Associate Professor in Economics for many years and it was from this that he made his incredible discovery in physics in 1987. This discovery was the turning point in his life and became the catalyst which inspired him to write a book on the subject.

Much of Poramest's writing career has centred around research papers and he is hoping that his book will be the inspiration that people need to become involved in science and the theories surrounding time. He is also planning another book called Happiness Economics.

Amongst many honours bestowed upon him, Poramest is the recipient of **The Most Noble Order of the Crown of Thailand** from H.M King Rama IX of Thailand (2012) and **The Most Exalted Order of the White Elephant,** also from H.M King Rama IX of Thailand (2015).

When he isn't working, Poramest loves nothing more than spending some quality time with his family, shopping or eating out with them. He also relaxes by practising Muay Thai for exercise, and devotes some of his time to teaching English for Thai people free of charge.

Poramest sees every obstacle as a stepping stone to success and it is his stated intention to stand present with peaceful mind as he goes towards the future.

You can contact Poramest Boonsri at;

Twitter; "Dr. Poramest Boonsri"

email: untolddiscovery@gmail.com;

marichatporamest@gmail.com

Indexes

Made in United States
Orlando, FL
04 November 2026

90207831R00203